Soviet Oil and Security Interests in the Barents Sea

Soviet Oil and Security Interests in the Barents Sea

Helge Ole Bergesen, Arild Moe and Willy Østreng

Frances Pinter (Publishers), London

First Published in Great Britain in 1987 by
Frances Pinter (Publishers), Limited
25 Floral Street, London WC2E 9DS

British Library Cataloguing in Publication Data

Bergesen, Helge Ole
Soviet oil and security interests in the Barents Sea.
1. Offshore oil industry—Government policy—Soviet Union 2. Offshore oil industry—Barents Sea
I. Title II. Moe, Arild III. Østreng, Willy
IV. Fridtjof Nansen Institute
338.2′7282′0916324 HD9575.S652
ISBN 0-86187-689-X

Typeset by Joshua Associates Limited, Oxford
Printed in Great Britain by Biddles of Guildford Ltd.

Contents

List of tables and figures vii
List of Abbreviations ix
Preface xi
Introduction xiii

1. The Barents Sea as an energy option 1
The development of the Soviet oil industry 1
The foreign trade factor 4
Energy consumption in the Soviet Union 12
Soviet investment policy 16
Alternative areas of future oil production 20
Notes 28

2. Soviet oil exploration in the Barents Sea 32
Seismic surveys 32
Exploratory drilling 33
Input of resources in offshore exploration 36
Soviet offshore activity: A preliminary evaluation 45
Conclusions 51
Future prospects 53
Notes 54

3. Petroleum and military strategy in the Barents Sea: A conflict of interests? 57
Ice conditions in the Barents Sea 58
The Northern Fleet: size, tasks and functions 61
Operational needs and conditions of the surface fleet 64
Submarines and petroleum activity 68
Transit conditions in the Barents Sea 79

SSN-operations in the Barents Sea 82
Summary and conclusion 83
Notes 85

4. Conclusions: Soviet options in the Barents Sea 88
Major alternatives 88
The Rouble Stategy 90
The hard-currency strategy 92
The form of cooperation 95
Notes 103

Appendix 1 The organization of the Barents Sea offshore research, planning, exploration and development 106

Appendix 2 Soviet offshore-related purchases from abroad 126

Index 139

List of Tables & Figures

Tables

1.1 Soviet investment growth, total and in energy sector 1970–85 and plan for 1986 17
1.2 Soviet oil production areas and reserves 21

2.1 Soviet mobile offshore drilling rigs 38

Figures

1.1 Soviet options 2
1.2 Petroliferous basins in the USSR 3
1.3 Soviet oil production 1975–85 5
1.4 Soviet oil exports to the West and the CMEA 1975–85. 6
1.5 Share of Soviet hard currency income 1973–84 represented by oil 7
1.6 Soviet hard currency income from oil 1973–84 in absolute terms 8
1.7 Major components in Soviet imports for hard currency 1982–4 8
1.8 Investments in the Soviet energy sector 1970–85 and plans for 1986 17

2.1 Exploration areas in the Barents Sea 34
2.2 The Barents Sea offshore organization 42
2.3 Norwegian supply bases 48

3.1 Maximum extension of ice in January 59
3.2 Maximum extension of ice in September 60

3.3 Ice drift direction and speed in the Barents Sea 60
3.4 Bottom topography of the Barents Sea 61
3.5 The naval base and military airfield concentration on the Kola Peninsula 62
3.6 The 200 metre depth contour 73
3.7 The marginal zone of the SSBN-stationing area in the Barents Sea 74
3.8 The year-round stationing area in the Barents Sea 75
3.9 Cross section of depths between Svalbard, Bear Island and North Norway 80
3.10 Internal and external transit routes in the Barents Sea 80

List of Abbreviations

ASW = Anti submarine warfare
Bcm = Billion cubic metres
dwt = deadweight ton
gos = gosudarstvenny, state (prefix) e.g. GosNIPI
LNG = Liquid natural gas
NGL = Natural gas liquids or gas condensate
mbd = million barrels per day
mbdoe = million barrels per day of oil equivalent
MW = megawatt, one thousand kilowatt
mt = million (metric) tons
NII = Nauchno-issledovatelski institut, scientific research institute
NIPI = Nauchno-issledovatelski proizvodstvenny institut, Scientific research and design institute
NPO = Nauchno-proizvodstevennoye obyedinenie, scientific production association
NTTs = Nauchno-tekhnicheski tsentr, scientific-technological center
SSBN = Ballistic-missile submarine, nuclear (strategic submarine)
SSN = Submarine, nuclear
VMNPO = Vsesoyuznoye mezhotraslevoye nauchno-proizvodstvennoye obyedinenie, all-union interbranch scientific production association
VNIPI = Vsesoyuzny nauchno-issledovatelski proizvodstvenny institut, all union scientific research and design institute
VO = Vsesoyuznoye obyedinenie, all union association (about enterprises)
VPO = Vsesoyuznoye promyshlennoye obyedinenie, all-union industrial association

Preface and Acknowledgements

This study brings together two major areas of interest to the Fridtjof Nansen Institute: Arctic security issues and Soviet energy policy. In order to understand the prospects for oil development in the Barents Sea, we have pooled our resources of knowledge from these two issue areas—let us hope in combination with a synergetic effect.

Where possible, we have used public sources, and as demonstrated by the references, most of the data in this study can be found scattered in Western and Soviet publications. In some instances there is no documentation or references because our sources have preferred to remain anonymous.

We would like to thank Jan Vanous and Marcia Levenson of PlanEcon Inc, Washington D.C., and Egil Bergsager for valuable discussions. We are also grateful for comments from Reidar Skarlo of the Norwegian Navy, M. O. Orning, Deputy of CINCNORTH, Tomas Ries of the Norwegian Institute of International Affairs, Anders Kjølberg, the Norwegian Defence Research Institute and Brit Fløistad of the Nansen Institute. Anders Solheim of the Norwegian Polar Institute provided crucial support by coordinating calculation of depth contours. Bernt Bull, adviser to the Minister of Foreign Affairs, Rolf Erik Rolfsen, chairman, Boconor and Nils Morten Udgaard, editor, *Aftenposten* gave valuable comments to a draft version of the book at a seminar organized by the Fridtjof Nansen Institute.

In our work on this book we have also benefited greatly from conversations with experts within the oil industry. Since most of these contacts prefer anonymity, we have refrained from thanking any of them by name.

In addition, we acknowledge financial support from the Norwegian Ministry of Foreign Affairs, both for this study and for a travel grant, which enabled a research trip to Moscow in January 1986, but the Ministry, in agreement with the authors, takes no responsibility for the contents of this volume.

We acknowledge practical assistance from Truls Andreassen, Erik Birch-Aune and Anne-Christine Thestrup in the completion of the manuscript.

Finally, we are very grateful to Rune Castberg for excellent research assistance. Without his perseverance and devotion to detail, this book would have been seriously lacking in documentation. However, the factual statements and the judgements in this book are the responsibility of the authors.

Helge Ole Bergesen
Arild Moe
Willy Østreng

Polhøgda, Oslo, September 1986

Introduction

The main purpose of this study is to analyse the prospects for Soviet petroleum production in the Barents Sea. We proceed from two rather different points of departure: Soviet energy policy, and military interests and needs in the area.

For various reasons, problems in the Soviet energy sector have become increasingly apparent. We shall review developments in the Soviet petroleum industry and discuss the implications for the Barents Sea.

The Barents Sea area has long been a haven for one of the most powerful military forces in the world. This is the Soviet Northern Fleet, crucial to the military strategy and superpower status of the USSR. As a consequence, the Barents Sea has been marked by military activity and East–West tension ever since the Second World War, and today these remote waters remain one of the core areas in superpower rivalry. Up to now civilian cooperation in maritime transport and fisheries has not been severely affected by military concerns, but in the early 1980s oil exploration in both the Soviet and Norwegian sectors of the Barents Sea introduced a new element into this picture. Such civilian activity has considerable economic importance and enormous growth potential, and could in fact change the constellation of interests radically on both sides.

With these points of departure, our aim is to provide answers to two questions: How important is the Barents Sea as an energy option for the Soviet Union? What economic, military and political problems will a possible development of petroleum resources in the area encounter?

Chapter 1 describes recent developments in Soviet energy production and policy, and outlines the major options now facing Soviet decision-makers. Since oil exports form the backbone of Soviet foreign trade, any analysis of the country's energy problems must pay close attention to its foreign economy. In the same chapter we therefore review the major alternatives for the USSR in this respect, with particular focus on prospects for oil exports to the West. These two sets of options, summarized below in figure 1.1 converge in one particular area: oil production. Thus the next step in our analysis is to evaluate major alternatives open to the Soviet oil industry in the 1990s, including the Barents Sea.

Chapter 2 deals with the development of Soviet oil exploration in the Barents Sea since the late 1970s. We give an account of activities to date and the equipment in use, with particular reference to foreign inputs. We compare the Soviet experience with a parallel development on the Norwegian northern shelf, and assess on this basis the future prospects for Soviet oil exploration in the area. This in turn leads us to the options sketched in figure 1.1.

Unlike most other offshore areas in the world, the Barents Sea has a dominant strategic element. Soviet military interests in both parts of the Barents Sea are analysed in Chapter 3, taking into account the present build-up of the Northern Fleet and the emerging American forward-based naval strategy. The purpose of this analysis is to pinpoint in concrete terms the possible conflict at present and in the future between the oil industry and the Soviet Navy. Starting with an examination of the present operational conditions for the Northern Fleet, we consider how these would be affected by future extension of petroleum activities. Furthermore, we discuss how the presumed tasks of the fleet may be influenced by such a development.

Chapter 4 is devoted to a discussion of the trade-off between military and civilian interests in the Barents Sea and the options for oil development in the area. As indicated in figure 1.1 they can be regarded as two partly overlapping possibilities: a rouble and a hard-currency alternative. As the prospects for large-scale compensation deals with Western companies are particularly interesting, we analyse the obstacles facing such an option, especially the military reservations, and the preconditions necessary for it to be chosen. We conclude that this option is possible, but not the most likely course of action at present. Under certain conditions, however, this may change rapidly and with little advance notice. Soviet energy production and exports have continuously produced surprises for Western observers, and this may apply to the Barents Sea as well. Consequently, our study contains few clear-cut conclusions, and no unconditional predictions. Our ambitions are limited to improving Western understanding of the relationship between different Soviet interests and the factors likely to influence Soviet decision-makers in this area.

In the final paragraphs we evaluate briefly how Soviet policy may affect Western interests, and how different Western actors may react to future moves from the USSR. This assessment is, however, limited to the field of petroleum cooperation. It therefore excludes the possible effects of offshore exploration and production on the pressing political and strategic issues in the area.

1 The Barents Sea as an energy option

The purpose of this chapter is to put the Barents Sea into the context of Soviet energy policy. To do this we must review the development of the Soviet oil industry as well as its current problems and evaluate the various options open to Soviet planners and policy-makers in the light of today's economic situation.

The development of the Soviet oil industry

Oil extraction in the Soviet Union started in the Caucasus towards the end of the nineteenth century. Baku, by the Caspian Sea, became the centre of oil activities. This dominance continued until after the Second World War, when large oilfields in the Volga–Ural region were developed. The USSR became a net oil exporter in 1951. However, coal was still the dominant fuel in the country, accounting for 66 per cent of total fuel production in 1951. Only from the mid-1950s is it accurate to speak of a deliberate oil strategy, and in fact oil did not surpass coal as the major fuel until 1968.

Oil production had been moving westwards up to 1965. In that year production in the European part of the country represented 72 per cent of the total. This pattern changed with the advent of the 'Siberian period': the first oil discoveries in Western Siberia were made in 1960. By the late 1960s production drilling was under way, and during the 1970s this area acquired the leading position among the Soviet oil producing regions. It now supplies about 60 per cent of the country's oil.

Since the late 1970s there has been growing concern about the prospects of this extremely important area. Up to the Siberian period, the Soviet oil industry had been able to move into new areas when needs outgrew the potential of existing producing regions. However, such a solution to today's problem is far more dubious.

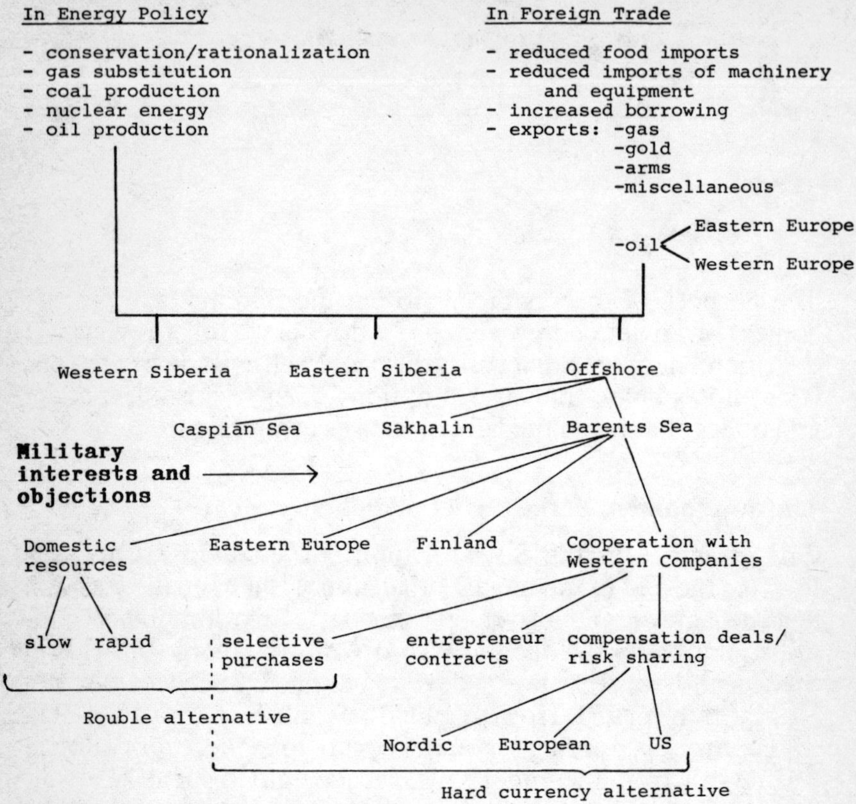

Figure 1.1 Soviet options (see Introduction)

Soviet optimism about the future of their oil industry after the breakthrough in Western Siberia was largely shared by Western analysts.[1] This 'idyllic' state was, however, shattered by CIA analyses from 1977, which in the two published reports[2] concluded:

- Oil reserves of the USSR were much smaller than previously estimated;
- Soviet oil production would peak within a few years and thereupon rapidly decrease;
- Soviet oil exports would cease by 1985 and the USSR would become a net importer;

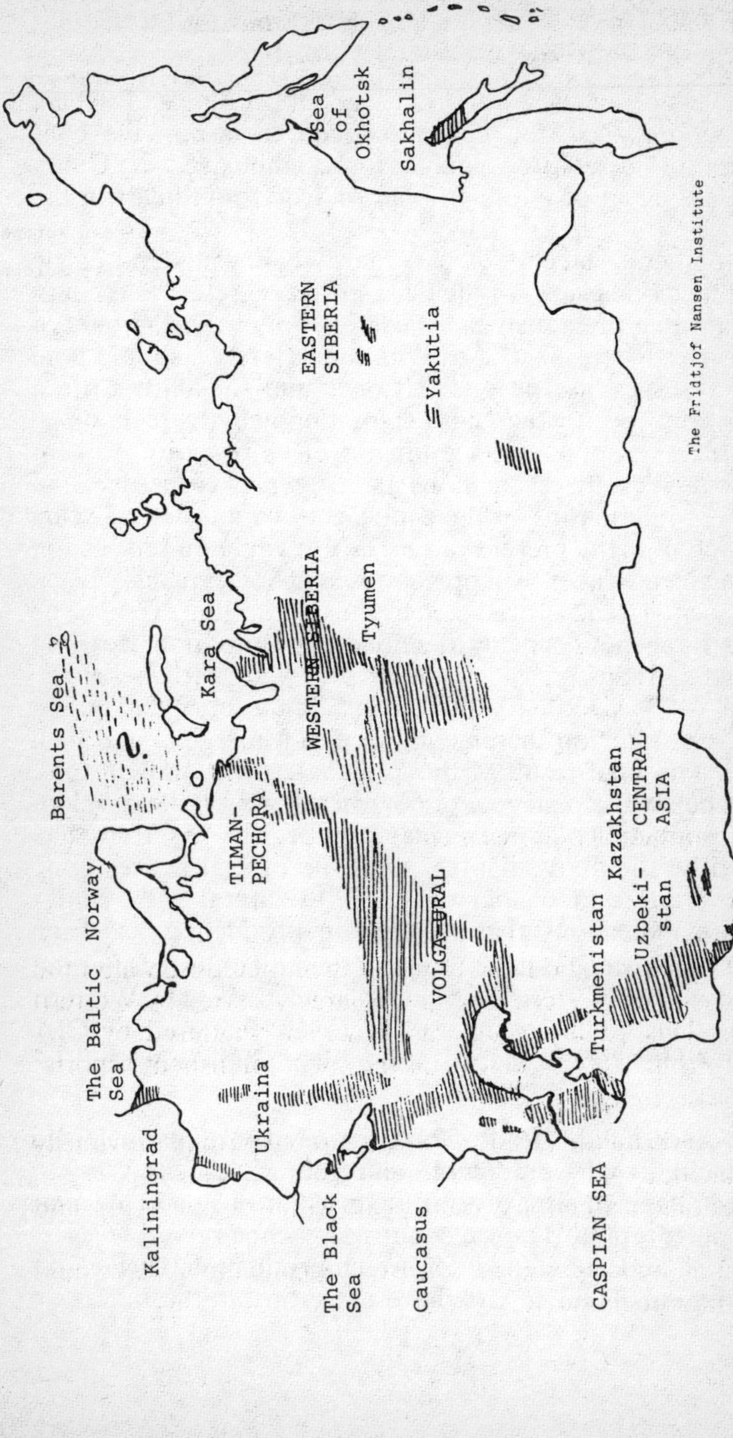

Figure 1.2 Petroliferous basins in the USSR

- The Soviet Union would not be able to replace oil with other sources of energy to a sufficient degree.

The publication of these reports triggered off a heated debate about the prospects for Soviet oil, and their possible consequences for Soviet foreign policy. For example: a Soviet Union in desperate need of oil might plan to acquire oil fields in the Middle East.

Developments since 1977 have partly confirmed and partly rejected the conclusions of the CIA. Soviet oil reserves seem to be smaller than previously believed, due both to the amount of extraction and to neglect of exploration. (Here we should note that 'oil reserves' are never fixed once and for all, but are a dynamic concept moving with exploration activity, technology and the price of oil.) Soviet estimates of its reserves are kept secret, but Western evaluations indicate proven reserves of about 6–11 billion tons, or the equivalent of about 10–18 years of production at the present rate.[3] The growth in oil production has been much weaker than presupposed in Soviet plans, while it has not been as weak as predicted by the CIA, although it must be said that the agency correctly identified some important trends in the Soviet oil industry. The result in 1980 was clearly below the plan goal (603 vs. 620–40 million tons). In the 11th plan period (1981–5) production increments were first marginal and then negative. The final result of this period was 595 million tons, (mt.) far below the plan goal of 630 mt. However, the development in production of crude oil may have been more serious than indicated by the above figures, since the official figures lump together crude and condensate (NGL—Natural gas liquids). There are indications that condensate production has been growing.[4]

The foreign trade factor

Soviet oil exports

On the other hand, the development of exports has differed greatly from the CIA predictions and from what many Western observers expected after production problems became obvious. Soviet oil exports to Western countries reached record levels in 1982, 1983 and 1984—years in which production levelled out and decreased. Total exports have not grown at the same rate

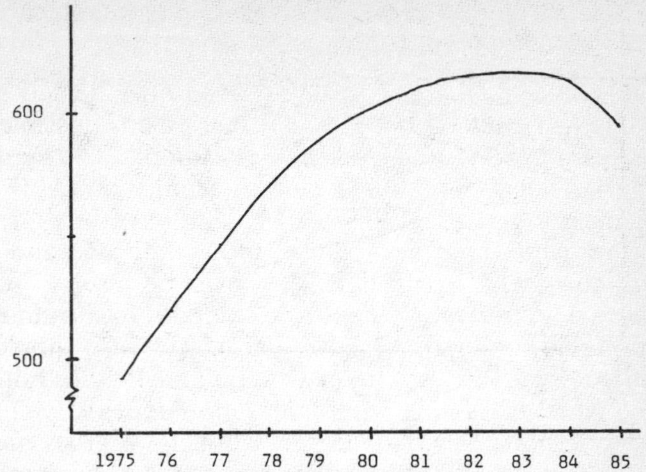

Figure 1.3 Soviet oil production 1975–85 including gas condensate (in million tons)

Note: y-axis has been cut to exaggerate the tendency
Source: Soviet statistics

since the USSR checked the growth in deliveries to Eastern Europe from 1980 and cut exports to that region by approximately 10 per cent from 1982. Still, total exports continued to grow until 1985.

This apparent inconsistency between production and export performance can be explained in several ways. First of all, we should keep in mind that oil exports to Western countries account for only about 13 per cent of total production (1983). (Total exports are about 30 per cent.) Consequently, if 1 per cent of the domestic consumption is conserved, this amount can boost exports to the West by 7 per cent. Furthermore, the supply of 'Soviet' oil for export is somewhat greater than revealed in production statistics, since the USSR has been bartering weapons for oil with countries in the Middle East for several years now. Estimates from PlanEcon indicate that record levels were reached in 1983 and 1984 with 12.5 and 14 million tons. In 1985 there was a slight decline to 13 mt.[5] But even if these volumes are far from insignificant, they cannot explain all the increase in exports. Apparently, domestic oil consumption peaked in 1982.[6] It is clear that the USSR has to a certain extent been able to conserve and find substitutes for oil. (See section on energy consumption in the Soviet Union.)

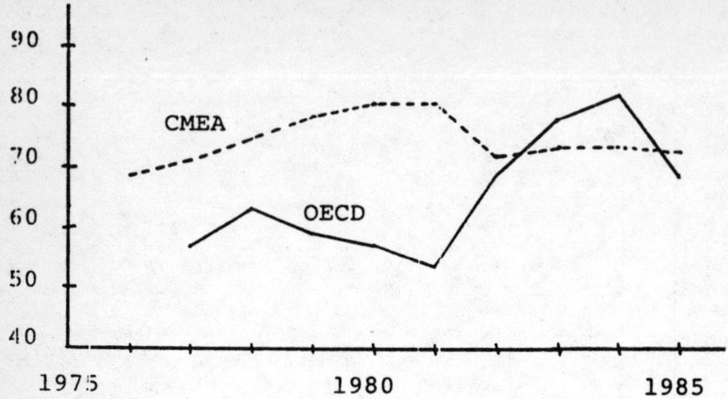

Figure 1.4 Soviet oil exports to the West and the CMEA 1975–85 (mt)

Sources: OECD countries: IEA/OECD: *Quarterly Oil Statistics*
CMEA countries: Directorate of Intelligence: *International Energy Statistical Review*

In 1985 the relationship between production and exports changed. Oil production fell by 2.9 per cent compared with 1984, but oil exports to the West dropped by approximately 12 per cent. Thus the decrease in exports was disproportionately high. The explanation seems to be the extraordinarily severe winter with increased heating requirements, in addition to general problems in energy conservation. All the same, there has evidently been a very strong effort to sustain oil exports, given the pivotal role of this commodity in Soviet foreign trade.

Soviet hard currency needs

If the Soviet Union were free to cut its oil exports, the domestic production problems could be tackled easily. But this is not a feasible solution. The country has become increasingly dependent on the revenues from oil exports in its foreign trade. Since the early 1980s, oil has provided 60 per cent or more of Soviet income in hard currency (not counting gold or weapons). (See figure 1.5.) The country's 'breakthrough' as an oil exporter came around 1970, and the oil revenues increased with soaring world market prices throughout the 1970s. Soviet imports have risen correspondingly. Although Soviet foreign trade seems to have been export-driven, this does not mean that import levels

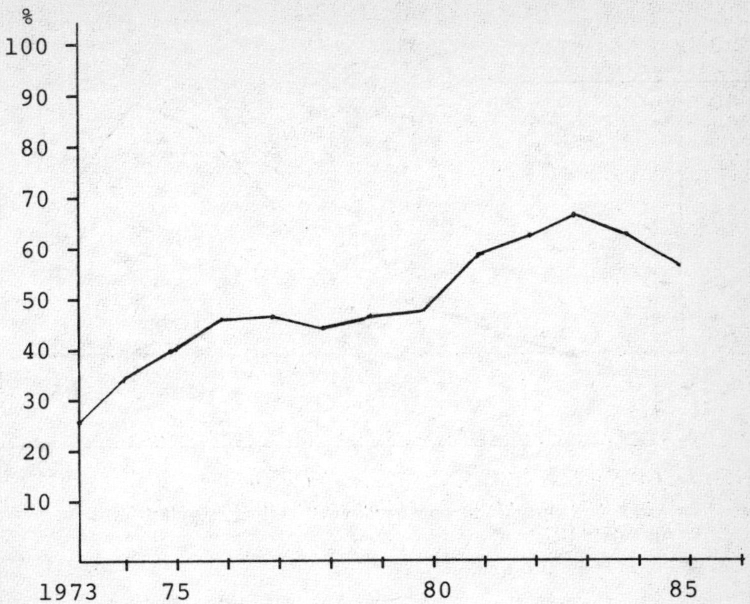

Figure 1.5 Share of Soviet hard currency income 1973–85 represented by oil

Source: Jonathan P. Stern, *Soviet Oil and Gas Exports to the West*, Aldershot, Gower, forthcoming 1987

can easily be adjusted if there are problems on the export side. Food, particularly grain, constitutes a large share of Soviet imports from the West. The need for imports will vary from year to year with the harvest, but over the last fifteen years the tendency has been increasing imports. One explanation lies in deep structural problems in Soviet agriculture, stemming from collectivization and poorly developed infrastructure. Increased consumption and more use of meat have also raised demand for grain.

The Soviet government could change the diet and reduce meat consumption without risking hunger in the population. Likewise, it could adjust prices to make them more 'realistic'. By doing so, waste and overconsumption would be reduced. But this policy would not be without problems. Such a change in diet would in itself be a disincentive for harder work, which is crucial to the present economic plans. On the other hand, if such a policy means a more rational distribution of food, it could contribute to

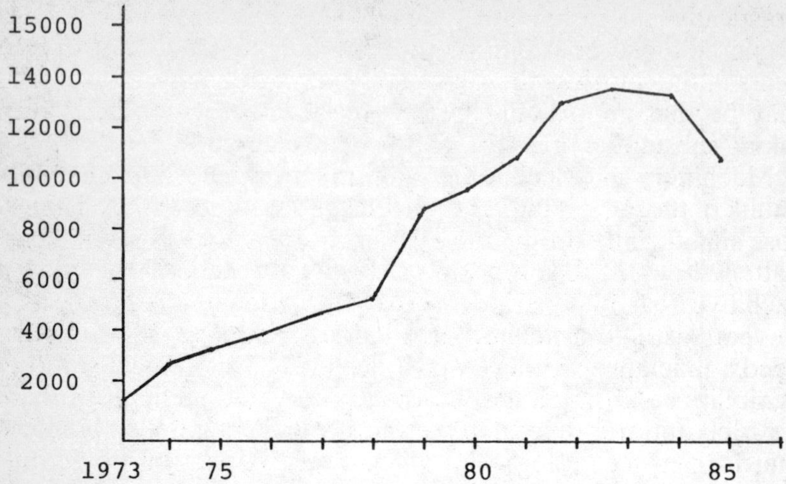

Figure 1.6 Soviet hard currency income from oil 1973–85 in absolute terms (thousand US $)

Source: Jonathan P. Stern, op. cit.

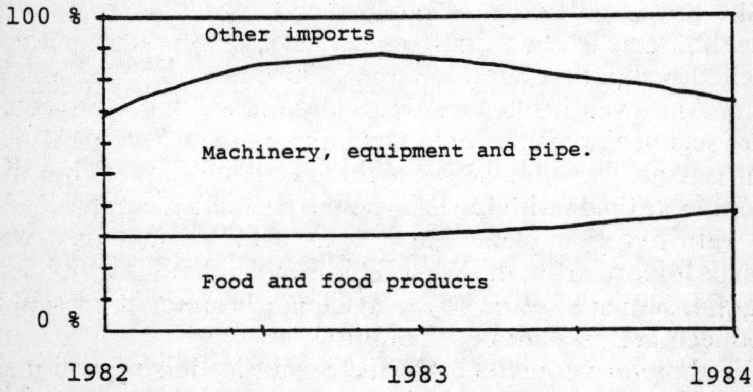

Figure 1.7 Major components in Soviet imports for hard currency 1982–4*

Source: *Vnesnyaya Torgovlya SSSR*, 1983 and 1984

* Including Finland

better availability of food and enhance the function of money as a stimulus to harder work. At the same time, prospects for increased agricultural output seem remote. Apparently, investments will not increase significantly in the coming years and there are no proposals for reforms that could change the picture drastically. Consequently, the Soviet authorities must be prepared to pay considerable amounts for imported grain in bad years in the future as they have done in recent years. However it may be that imports in 'normal' years can be abolished, thus taking substantial pressure off Soviet foreign trade.

Machinery and technology imports comprise the other big chunk of the import bill. The hard-currency import of such items rose significantly during the 1970s, but recently this growth has flattened out. (There is always a downward trend at the end of each five-year-plan period, due to completion of large projects.) In years with a particularly bad harvest, and large food import needs, machinery imports have declined. This indicates more flexibility regarding imports of machinery and technology than for food imports, at least in the short term. But it is also obvious that foreign technology had a significant impact on Soviet industry in the 1970s, especially in the chemical, energy, and automobile industries.

In the Soviet Union today there are fewer than before who believe in Western technology as a universal remedy for the ills of the Soviet economy. On the other hand, it seems that the growth targets of the 12th five-year plan (1986–90) cannot be reached without sustaining the present import level as a minimum. And even if the attitude to large new import projects is more sceptical, portions of Soviet industry have become dependent on the West for spare parts and extensions of installed equipment. Trade with East European partners is emphasized in the 12th five-year plan,[7] but it is hard to see that this could replace imports from the West. This is quite obvious with regard to grain; similarly, some of the machinery cannot at present be produced in East European countries.

We therefore conclude that the Soviet leadership will make substantial efforts to maintain hard currency earnings at the present level to pay for imports. In the following we survey the alternatives they face in this respect.

Soviet acquisition of hard currency

As mentioned, oil is easily the major export item. The question is whether other export items can replace oil, at least partly, as hard-currency earners.

Gas Gas exports to Western countries have been growing, but in 1983 constituted no more than 12.3 per cent and in 1984 16.2 per cent of value of exports to the West.[8] Only a few years ago it was widely held that in the late 1980s gas would at least equal oil as a hard-currency earner. This prediction has been proved wrong, for two reasons. Firstly, the expansion of the West European gas market has not been as rapid as anticipated, so the need for imports has been smaller. Secondly, the price development has not been as positive (from a seller's point of view) as anticipated. A significant jump in imports that would affect the USSR cannot now be expected before the late 1990s, at the earliest. And even then, actual revenues will probably be lower than originally expected for the late 1980s. Needless to say, low oil prices will have repercussions on the price of gas.

Gold Gold exports are not made public and therefore not included in Soviet foreign trade statistics. The reserve position of the USSR is, however, very good, estimated as being $25 billion at the end of 1984. The value of gold exports was estimated as $750 million in 1983 and $600 million in 1984.[9] Annual production is thought to be 300–30 tons and stocks are increasing.[10] Exports could undoubtedly be increased significantly in the short term to ease the foreign trade situation. However, increased Soviet exports would press the gold price down, so this does not seem a viable long-term solution. (The risk of a continued fall in gold prices is particularly imminent now, since the world's other major gold producer, South Africa, is in an even more difficult foreign trade situation.)

Arms Soviet exports of arms have been estimated at approximately $8 million in 1983, $7.3 million in 1984 and $5.5 million in 1985.[11] Most of this goes to countries in the Middle East on credit terms, but ultimately arms exports are paid for in crude oil. Sales seem to have slumped with the deteriorating economies in many Middle Eastern countries, together with lower levels of destruction in on-going conflicts in the region. Their present economic situation makes it unlikely that the oil exporting countries in the Middle East will increase the level of arms imports, except for Libya. In any event, arms may not be a very attractive export item from a Soviet point of view. Sophisticated weaponry includes complicated machinery and scarce metals that could have an important alternative use in the domestic economy. It must be kept in mind, however, that the Soviet government also has political reasons for maintaining arms exports.

Increased borrowing The USSR's credit-worthiness is still considered good, and so the country can solve short-term problems by increased borrowing in the West. If prospects for future export earnings diminish, Western banks may become less interested in lending to the USSR. East–West political relations will also affect the readiness of Western governments to guarantee loans. Unlike most East European countries, the USSR has traditionally been highly conservative in its borrowing policy, and we believe this will be the general principle in the future as well. In the beginning of 1985 the USSR had to draw down its assets in Western banks to a significant degree. Later the same year borrowing increased, to rebuild deposits. Net debt to the West increased by $4.3 billion in 1985, making up a total of $9.6 billion at the beginning of 1986.[12]

Other export items The USSR has for a number of years pressed for a broader portfolio of export items, without much result. Various metals and minerals, paper and wood and some machinery constitute the majority of such exports for hard currency. The export outlook for these commodities does not seem promising enough for them to replace part of the oil revenues, however.

Eastern Europe Some of the problems the Soviet Union is experiencing in its economic relations with Western countries are likely to be passed on to Eastern Europe. Oil deliveries will probably be cut; Eastern Europe will be pressured to substitute more Soviet gas for oil, conserve more energy and buy some oil from Middle Eastern countries for hard currency or on barter terms. This will affect the economic development in Eastern Europe adversely and hurt its foreign trade balance *vis-à-vis* the West. Consequently, the Soviet Union must be aware of the political costs it could incur. The potential gains for the Soviet Union will thus be limited. A 25 per cent cut in oil supplies to the area (except Poland, which would have severe problems absorbing such a sizable cut) must be considered drastic but not improbable. This would free 13–14 million tons for hard currency exports or the equivalent of 21 per cent of 1985 export levels to the West.[13]

Thus, it should be clear that maintenance or increase of oil exports is the key to sustaining import levels essential for the Soviet Union's current economic plans and political priorities. Furthermore, we argue, a falling world market price for oil is not an obstacle to exploration—contrary to the tendency elsewhere. The reason is the lack of alternatives in Soviet exports. Falling prices necessitate exporting even more oil. It is too early to tell exactly what domestic sacrifices the Soviet government is willing to bear for this purpose. The need to sacrifice depends largely on developments in domestic oil consumption.

Energy consumption in the Soviet Union

Data on Soviet energy consumption are sparse, but they suffice to show that energy requirements are relatively higher than in Western industrialized countries. Western estimates of unit energy requirements indicate that these have increased during the 11th five year plan, whereas they have fallen significantly in most Western countries.[14] It has been estimated that the USSR consumed 33 per cent more energy per unit GNP than the United States in 1983.[15]

For some years the conventional wisdom among Soviet planners has been that it would cost far less in terms of investments to conserve an additional barrel of oil than to produce it in remote and costly fields in Western Siberia. The costs of conservation could probably be as low as one third of those involved in production.[16] Against this background, some reallocation of resources might be expected—investing in substitution and energy saving instead of spending the money on increasing the oil output at ever-increasing costs. The need for stronger conservation measures has long been a theme in Soviet newspapers; the issue was treated on the first pages of the published parts of the long-term energy programme of 1984.[17] However, little has actually been done in the field of energy conservation. This can probably be explained in terms of general development strategies in the USSR: namely extensive versus intensive growth.

The development of the Soviet economy has been characterized mainly by extensive growth: increasing ouput by means of increasing inputs, such as capital, manpower and natural resources. The Soviet economic system has largely been designed around this strategy, so the 'extensive growth alternative' is the natural preference of the system. But when input factors become scarce, as is the case in the USSR today, intensive growth becomes essential: i.e. increasing the productivity of these inputs. For many years now, the Soviet leadership has emphasized the need to shift to intensive growth. As indicated above, however, this strategy does not fit the basic design of the Soviet economic system, based on central planning and directives. It requires local adaption, for which there is little leeway and few incentives.

Moreover, the Soviet system lacks what in a Western economic system could be termed 'rational indicators for success and failure'. Since cost analysis is poorly developed, optimization of costs and output is difficult to attain, and the system is left with physical quantities as the paramount indicator for economic growth. This indicator also fits nicely into a planning system based on physical units. Consequently, the system presupposes eternal growth of output and is not prepared to encounter reduced output of any resource. (This attitude fits manufacturing industry better, where the general rule is decreasing marginal costs; in extractive industries, such as the oil industry, marginal costs must at some point rise.) The growth expectations are highlighted in the oil industry, because of the strategic significance of oil and because so much political prestige is attached to it.

One general problem of energy consumption in the USSR is that the Soviet economic system prevents a more economical utilization of energy at the enterprise level. The central allocation of inputs to industry, combined with a very weak connection between production costs and energy prices, gives the managers few incentives for energy conservation. The same arguments apply to energy substitution - switching from one fuel to another, e.g. from scarce oil to abundant gas.

Theoretically one could conceive of means to induce the different entities in the economy to use energy more economically. One such remedy could be to introduce prices that reflect the marginal costs in oil production. The concept 'marginal costs' is, however, rejected in principle, as Soviet price formation is based on average production costs. Oil is therefore still relatively cheap. But even a drastic increase in prices would not necessarily change the behaviour of enterprises, unless they were given more independence in choice of inputs. It is important to remember that in the Soviet system prices do not have the same allocative function as in a market economy. Soviet managers are not free to choose inputs—say, gas instead of oil—even if they realize that the latter is more expensive, unless the enterprise is scheduled to change to gas in the plan. In fact, the problem concerns more than direct fuel consumption by enterprises, since most energy is wasted through overconsumption of energy-intensive materials, such as steel, where prices do not reflect the high and costly energy content. Still, even if managers lack the freedom to change inputs, they do have the opportunity to save on any given resource. It goes beyond the scope of this book to discuss further the incentives that exist to facilitate such energy savings.

The most promising area for significant savings of oil seems to be large power stations where oil can be replaced by natural gas. At least theoretically, the planning system should be well suited for such decisions. Soviet calculations show a potential for saving 1.5 million barrels per day (75 million tons a year) in the power generating industry from the 1980 level, without major investments.[18] But even if this is technologically feasible, important institutional barriers slow the development of a gas grid to supply the power stations and the building of gas storage facilities.[19]

In sum: Soviet conservation performance has not been good so far, nor do efforts seem to have been that strong. Until recently it has been relatively easy to cover increased demand for energy, and particularly oil, by increasing output, the traditional Soviet response. Only gradually has attention been turned to the demand side itself. Still, the squeeze has not been made hard enough. This may now be changing, and consequently past performance has limited predictive value. In other words in a desperate situation, far more oil can and must be conserved, but it is a big step from insight to action.

The development of the nuclear power programme is another factor influencing the consumption of oil. The Soviet Union still has less installed nuclear power capacity than France, and only a third of that of the United States,[20] but output is growing at a rapid rate. In 1985 nuclear electricity constituted 11 per cent of total electricity production. According to the present five year plan, the goal is to produce 390 billion kilowatt-hours in 1990, some 21 per cent of planned electricity output that year. A natural question is how the accident at the Chernobyl nuclear power station will affect the Soviet energy situation, and Soviet energy policy.

According to the Soviet officials reporting on Chernobyl in Vienna in August 1986, two of the four reactors at Chernobyl will be reopened in 1986, the reopening of the third is uncertain, and the fourth, in which the accident took place, will be buried in concrete. Taking this statement at face value implies an immediate loss of 4000 megawatt capacity this year. In addition there is the fifth reactor, originally scheduled for opening in 1986, the state of which remains uncertain. The drop next year compared with the plans could be 2000 megawatts plus the fifth reactor, another 1000 megawatts. However, the planned increase in nuclear capacity in 1986 was 7500 megawatts, and the lost output of three reactors for one year only amounts to approximately 1.3 per cent of total electricity output in 1986.

But the accident will slow down the pace of the nuclear energy programme for several years. The 1990 goal is now clearly unattainable but even before the accident it was held to be unrealistic. The underfulfilment will have to be covered by other forms of primary energy if the planned growth in electric power output is to take place. The most likely replacements are coal and gas because production of these fuels is on the increase. But in a fairly integrated energy system the accident will also affect oil, although indirectly.

Chernobyl will affect the Soviet economy in general and the energy sector in particular in several ways: the immediate reduction in power generating capacity as well as the delay in developing the industry will put more pressure on other sources of energy. Cuts are likely to be shared between the three 'markets' for Soviet energy: domestic consumption, Eastern Europe, and Western importers of Soviet energy. Thus the accident also affects hard-currency income. If one third of the lost output of electricity this year (120 mbdoe: million barrels a day oil equivalent), is covered through reduced oil exports (compared to what they would have been without the accident), the loss in hard currency will equal $220 million at $15 per barrel.[21] This is a theoretical exercise, since exports in the real world have been growing for quite other reasons.

The radioactive contamination of parts of the Ukraine will necessitate increased imports of food, putting more pressure on Soviet foreign trade. It is too early to assess the magnitude of this increase. In addition to these immediate concerns, the accident will probably have long-term repercussions on the Soviet energy industry. In any case the replacement of the damaged reactor and cleaning the three others as well as one under construction will require capital, to be taken from somewhere. The whole rescue operation has been costly, and there are of course indirect effects on other parts of the Soviet economy. However, it is not logical that these costs will be covered by adjustments solely in the energy sector, let alone the nuclear power industry. It seems more probable that the costs of this emergency will be distributed more or less evenly throughout the Soviet economy.

In this limited economic view Chernobyl just adds another constraint to an already troubled economy. Neither the lost capacity, nor the replacement costs are big enough to necessitate new priorities in other parts of the energy sector. But it is not totally impossible that the accident will mean some reorientations in Soviet energy policy for political reasons by weakening the hand of the 'nuclear lobby', thought to have a stronghold in Gosplan, and supporting spokesmen for traditional fuels. So far this is very speculative, and it will take some time before major changes in the energy policy can be detected. To investigate further the Soviet options and the alternatives likely to be chosen by the Soviet leadership, we must take a brief look at Soviet investment policy.

Soviet investment policy

In the last half of the 1970s, total investments in the Soviet economy grew at an average rate of 3.3 per cent. In the first half of the 1980s (the 11th five-year plan) growth has been somewhat above this, 3.5 per cent. However, the growth rate in the energy sector has been consistently higher than total growth. The energy sector increased its share of investments rapidly in the 1970s, from 5.1 per cent of total investments in 1970 to 13.2 per cent by 1982. That year energy investments amounted to 26 per cent of industrial investments.[22]

Turning to the share of the different energy carriers, we see that the share represented by the oil industry has been growing

Table 1.1 Soviet investment growth, total and in energy sector 1970–1985 and plan for 1986

Branch	1970–75	1976–80	1981	1982	1983	1984	1985	1986
Total	6.9	3.3	3.5	3.6	5.7	1.8	3.0	7.6
Industry	6.7	3.5	3.9	2.8	5.5	3.0	3.9	18.0
Energy	5.8	6.8	8.7	6.6	5.7	4.2	5.4	26.0
Electric power	3.2	3.9	1.8	0.0	2.2	7.2	5.2	24.0
Coal mining	1.9	3.7	0.0	10.0	4.5	1.4	0.0	27.0
Oil extraction	8.6	12.3	18.8	7.4	4.6	−1.1	4.0	31.0
Gas extraction	11.3	2.8	0.6	15.0	17.4	18.4	13.5	15.0

Source: *PlanEcon Report* 1–2, 1986.

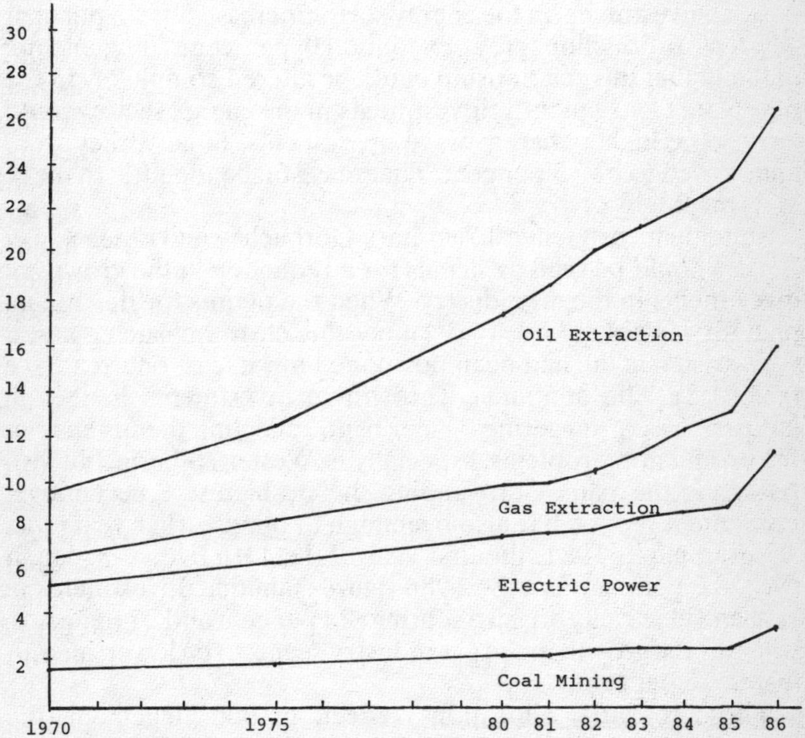

Figure 1.8 Investments in the Soviet energy sector 1970–85 and plans for 1986 (in billion roubles at estimate prices of 1 January 1984)

Source: *PlanEcon* Report 1–2, 1986

disproportionately. This development is due to soaring marginal costs in the oil industry, not only in new fields, but also in old ones where enhanced recovery methods are being introduced to recover more oil.

With this rather dramatic development in mind, we should ask if the time for change has arrived in the USSR. Energy, and particularly oil, is taking a growing part of the country's financial resources, to the detriment of other sectors. The new five year plan was surrounded by many expectations in this respect.

Investments in the 12th five year plan

The 12th five year plan (1986–90) calls for an accumulated increase in total investments of 18–22 per cent over the coming five years. The slated growth in energy production is the same, 18 per cent in 1990 as compared with 1985. In the 11th five-year period, investments in the energy sector increased by 35 per cent whereas production grew by only 10 per cent. It would be unlikely that this relationship could be altered completely in the new plan. Consequently, investments in the energy sector would have to be higher-than-average investments. Finally, they were announced to be 35 per cent. That may still be too little to make ends meet.[23]

Statements by General Secretary Gorbachëv and other Soviet leaders could be seen as signals for a reduction in the growth of investments in the oil industry. When the figures for the annual plan for 1986 were released[24] it became clear that such goals, if they existed at all, had been postponed for at least one year (see table 1.2). The announced growth in investments in the oil industry was a staggering 31 per cent, reflecting the urgency of the production problems, especially in Western Siberia, but also reflecting the panic surrounding the problems. True enough, investment growth has also been high in other earlier years too: for example in 1981, the first year of the 11th five year plan, it was 18.8 per cent, but the 1986 figure is unique. Investments in coal and electricity are also soaring (24 per cent and 27 per cent), whereas the growth in the gas industry seems to be lower, an estimated 15 per cent.

These resource allocations reflect, in our judgement, the immediate concerns in the Soviet energy sector. The favoured solution seems to be 'more of the same': increasing inputs even more. Priority is given to policies that can offer immediate relief.

This means a continued reliance on oil (since this requires less or no adaption in the rest of the economy), but also on coal and electricity. Oil production is still planned to increase, to 635 million tons in 1990, i.e. by some 6.7 per cent compared with 1985. In this politico- economic environment all long-term solutions suffer.

In the Soviet economy this has repercussions on long-term investments involving energy conservation and substitution policies. Thus, this short-term policy also affects the gas industry, which 'suffers' relative to other energy sources. In the gas industry, planned output growth is well below the growth in the last five-year period (some 30–32 per cent versus 48 per cent)[25]. The constraints do not lie in the extraction of gas, nor do the big trunklines seem to constitute a major obstacle towards increased utilization of natural gas. Rather, it is the refurbishing of plants and power stations and the construction of smaller distribution lines that slow down the gasification process. So, even if it is obvious that substitution of gas for oil would be a rational strategy, given the distribution of energy resources, such a shift is difficult to undertake because it would yield substantial results only after some time. Hence the old solution, maintenance or increase of oil output, is preferred because increased investments in the oil industry can have effects in the short term.

If this traditional policy is pursued, the most likely path to follow in the face of immediate problems is to intensify extraction in areas already in production, rather than putting resources into risky ventures in unexplored areas. Consequently, more resources are being channelled into the old fields, especially in Western Siberia.

This can also be seen as part of a general tendency towards risk aversion in the Soviet economy. Economic risk taking is not easily incorporated into a planned economy. Officials responsible for risky projects that fail will often have to pay a political price, in terms of reduced influence or even complete ousting. At the same time, rewards for successful projects are not correspondingly high. Realizing this, decision makers will not be inclined to advocate projects with uncertain prospects. Moreover, in times when there is a squeeze on the economy it will be even harder to argue for risk taking than when investment capital is more plentiful.

These developments point to the following simple guidelines

for action in Soviet energy policy: *quick results with as little risk as possible*. In the Soviet oil industry, this implies a continuation of the policy that has led to many of today's problems such as inadequate exploration and maintenance of production wells. As an effect of this policy, the reserves/production ratio has deteriorated and maintenance costs are soaring.[26]

The argument above has been presented in a somewhat simplified form for the sake of clarity. Taken literally, it would in fact mean that no such thing as oil exploration could take place in the USSR since risk is involved and results are distant. What we mean is that the system does not favour risk taking, and that in the present situation it is hard to find room for long-term investments. One possibility is exploration in areas where there already is production and the chances of finding new reserves are high. In such a case, exploration can be seen as an integrated part of the local oil industry, an activity necessary to maintain or increase production volume, and ultimately to secure the survival of the production organization. (On the other hand, we have seen in Western Siberia a strong tendency to divert resources away from exploration to production.) It is quite another thing to explore in new areas where there are no well established organizations that depend on new discoveries.

With this in mind, we will look at some of the more interesting areas for future oil production. Soviet oil production takes place in a number of regions, and many of these will continue to play an important role in the future. What we are interested in here is to focus on areas with a potential for a substantial increase in production.

Alternative areas of future oil production

Western Siberia

For a number of years now, this area has been responsible for net increments to total production, while the old provinces in the western part of the USSR have seen their output fall. In 1985 this pattern changed as West Siberian production fell compared to 1984. The problems in Western Siberia stem partly from depletion of reserves, but mainly from the production strategy or the lack thereof. Pressure for increased production has led to a neglect of exploration and maintenance. This is now commonplace criticism within the USSR generally.[27] The extraction method itself, with heavy reliance on water injection from an early stage, may also have damaged the fields, although this is disputed by most Soviet specialists.[28] The accumulated problems became obvious in 1985, triggering an emergency political response which included increased use of labour from other production regions, especially for maintenance work. Altogether 100 000 men from production organizations in other regions of the country are now being flown in for fifteen days' duties in Western Siberia.[29]

Table 1.2 Soviet oil production areas and reserves (mt)

Area	Accumulated prod. 1–1986	Prod. 1985	A + B reserves	Potential
1 Western Siberia	3 613	372	2 240	7 400
2 Volga–Ural	5 102	139	2 110	960
3 Caspian Sea, Central Asia, Caucasus	2 186	54	692	2 330
4 Timan–Pechora	189	18	705	550
5 Ukraina	179	6	27	140
6 Byelorussia	106	3	14	40
7 Sakhalin Island	85	3	3.3	140
8 Others	78	—	58	428
Subtotal	10 943	595	5 849.3	11 988
Outer continental shelf*			—	6 178
Total	11 538	595	5 849.3	18 166

* Barents, Kara, Laptev, East-Siberian, Chukchi, Bering and Okhotsk Seas.

Reserve data are not published in the Soviet Union. These reserve estimates are from 1980. 'A + B reserves' in Soviet terminology roughly parallels 'proven reserves' in Western. 'Potential' is a very vague category, covering all reserve estimates in addition to the proven reserves. Hence, these figures must only be taken as loose indications.

Source: E. N. Tiratsoo: *Oilfields of the World*, Beaconsfield, Scientific Press, 1984. *Soviet Geography*, various issues.

The future development in Western Siberia will be a key to evaluating what options the Soviet Union has in its oil policy. If a new giant oil field were to be found in that area, the Soviet leadership could buy some time. The infrastructure in the area is fairly well developed by Soviet standards and a new field could hook up to existing pipelines. However, there are no signs that such finds have been made recently. The question has been raised whether such a find could be kept secret, but in the present difficult situation, a large discovery would be developed as rapidly as possible—and probably with much publicity. In recent Soviet oil history there is rather a precedence for announcing a new major find too early. The Salym field in Western Siberia, discovered in 1965, was for some time rumoured to be the World's largest oilfield. However, production to date has only reached 5600 barrels per day.[30] But there is still belief in the Soviet Union that new major oil finds can be made in Western Siberia. One argument is that oil could be found further north under the huge gas fields and also elsewhere in deep horizons. Such opinions have been voiced particularly by members of the scientific community at the Academy of Sciences's branch in Novosibirsk.[31]

There is no doubt that new fields will be opened up in Western Siberia. Altogether more than 300 finds have been made, including twenty giants.[32] By the end of 1980, thirty-nine were in production. During the 11th five-year plan twenty-eight new fields were slated for production, fourteen in 1985 alone.[33] There is suspicion however, that some of the fields were 'producing' only in official reports and not in reality.[34]

In the current five-year period (1986–90) sixty-four new fields are to be put on stream according to plan,[35] making a total of 131 fields in production by 1990. The problem is not the number of fields, but their size and costs. According to one Western estimate, costs in the most expensive new fields in Western Siberia are approximately 250 roubles per ton, or thirty-four roubles per barrel! The *average* production costs in the Soviet oil industry are estimated to be twenty-six roubles per ton.[36] (The dollar equivalents at the present official exchange rate would be $48.5 and $7 per barrel.) But even if Western Siberia is costly, and increasingly so, there is a certain security in projects there that other areas cannot offer. The outlook for short-term pay-off is far more promising than elsewhere. This gives Western Siberia a strong lead in a situation where short-term solutions are preferred and risk is avoided. It is already clear that the region must play a pivotal role in balancing the total oil output in the 12th five year plan. Production is planned to increase by some 20–24 per cent in 1990 over 1985, i.e. 445–60 million tons.

Eastern Siberia

There is much uncertainty about the potential in the vast area to the east of the oil-rich provinces in Western Siberia. Again the

'Siberian lobby' is very optimistic. According to Abel Aganbegyan, former head of the Institute of Economics and Industrial Management in Novosibirsk, this region could be a major oil and gas producer by the year 2000[37], but this is disputed by others.[38] So far only gas production on the level of one billion cubic metres per year has taken place, in Jakutia.

Eastern Siberia's potential may be uncertain, but the costs are certain to be enormous, given the region's remoteness from the industrial centres of the country and the poor infrastructure. As yet there has been no breakthrough for this region, but it remains an option.

Kazakhstan

Kazakhstan, in Soviet Central Asia, has become the second most important oil producing Soviet republic after RSFSR (the Russian republic), with a production level of 22.8 million tons in 1985. Output is scheduled to increase to 30 million tons in 1990.[39] The most promising field in the area, the Tengiz field, displays many difficulties, and new infrastructure is required.

Offshore areas

Approximately 20 per cent of the world's total continental shelf belongs to the Soviet Union. According to Soviet sources most of it is potentially oil and gas bearing.[40] Much of it must, however, be regarded in a very long-term perspective. That goes for all of the shelf north of Siberia. Below we will look at the present or shorter-term projects.

The Caspian Sea Oil production at sea has taken place since the 1920s, in very shallow waters, partly from artificial islands. Production in that part of the Caspian Sea has been falling for some years. But new oil discoveries in the '28th April' field, made at water depths of 80–120 metres in 1979, were promising. To develop this new region it has been necessary to upgrade offshore technology significantly. Mobile rigs for exploration were introduced in 1980, and drilling is undertaken down to 6000 metres. Production in the 28th April field takes place from four fixed platforms at water depths between 84 and 110 metres with an annual output of about 3 million tons.[41] At present, apparently four semi-submersible rigs and seven jack-ups are involved in exploration in the whole Caspian Sea.[42]

The Caspian region (both onshore and offshore) is singled out in the 12th five-year plan where it is stated that the deep oil finds in the area are to be brought into production. It seems reasonable to assume that the input of resources to the region will increase in the years ahead. But *total* production from the region is planned to grow only some 10 per cent to 1990, since the new fields must compensate for a sharp fall in production from the older ones.[43] The *offshore* production, counting both the old shallow fields and the new deep ones, was estimated to 8.5 million tons in 1985, and will probably grow to 10–10.5 million tons in 1990. [44] Exploration efforts in the 12th five-year period are expected to be concentrated on the Kaverochkin field, next to the 28th April field, but there is still a lot of exploration to be done in other parts of the Caspian.

The Sea of Okhotsk (Sakhalin) The Sakhalin offshore project was developed in the mid-1970s as one way of securing oil supplies for Japan. Development was to take place on a joint basis. The Japanese partners in Sodeco (Sakhalin Oil and Gas Development Corporation) provided the USSR with a loan of $100 million for exploration. The loan was to be repaid in oil found. The fate of the project is not yet finally decided, but severe problems have emerged. First of all, very little oil has been found, although there is gas in ample quantities; 20 million tons of oil and 140 billion cubic metres of natural gas plus 13 million tons of gas condensate in the Chaivo field according to Soviet sources. The Odoptovskoye is believed to contain more oil, but probably not enough for exports of any scale. All the oil will be needed for domestic consumption in the Soviet Far East. For this reason the whole venture has been turned into a LNG (Liquid Natural Gas) project. The market for LNG in Japan is, however, more or less saturated and interest in deliveries from the USSR are at present very moderate, even if positive statements are still made concerning Japanese willingness to take Soviet gas.[45]

Consequently, the Sakhalin project cannot be considered a major alternative in oil terms. It is of interest as an example of cooperation with Western countries in the energy sphere and it must have provided the USSR with important technological experience. Even if it cannot compete in terms of reserves, an eventual development of gas could compete with other projects in terms of investment and political attention. Hence Sakhalin is not irrelevant in our context. At present the most important and uncertain factor is Japanese needs and priorities. Soviet domestic consumption may be expanded if the delayed pipeline project to Komsomolsk na Amur is completed.[46] So far the local needs are below the production level *onshore* at Sakhalin.[47] In any case there are plans for further offshore operations in the Far East. At present three jack-ups and one drill ship are active in the Sakhalin area, but the plan is to move exploration eastwards to the western shelf of Kamchatka and even further north to Anadyr.[48] Presumably, this will take some time.

Some political lessons from Sakhalin are also likely to have an impact on decisions regarding other projects, e.g. the Barents Sea. The unresolved Northern Territories' issue (the Kuriles) has been a significant factor in Japanese attitudes to cooperation with the Soviet Union, indicating that good political relations with major partners are close to a precondition for a successfull project of this kind. Secondly, the project proved vulnerable to American sanctions in 1981–2, because of the large component of American technology. The sanctions seriously reduced drilling capacity in 1982, from an expected five wells with an American drilling ship to two with a Soviet replacement vessel.[49] This lesson, a parallel to the pipeline conflict experience, can on the one hand serve as an argument against US partners; but on the other hand, if Soviet authorities realize that some dependence on American technology is unavoidable, it may also serve as an impetus for gaining some sort of US commitment to a project (cf. our discussion at the end of Chapter 4).

The Barents Sea The Barents Sea is another area with a potential for large oil finds, but so far no commercial discoveries have been made (see Chapter 2). According to a Soviet source, it is believed that 35 per cent of all expected offshore oil resources are accumulated in the European part of the USSR.[50] Soviet estimates from the 1970s indicated a potential of 2200 million tons of oil in this area.[51] This number was based on rather limited geological knowledge, but may explain some of the Soviet interest for starting exploration drilling in the area.

Developments in adjacent onshore areas are of interest. Oil and gas finds in the Timano–Pechoran province can be taken to indicate oil and gas reserves in the Pechora and Barents

Seas, as can also the small finds on Kolguev Island (see map, figure 2.1).

In the Timano–Pechoran oil and gas province (the Komi autonomous republic and the Nenets autonomous district, also loosely referred to as the European North), oil production is at present about 18 million tons a year. A number of new oil and gas finds have been made in recent years, of which several are located on the coast. The remoteness of these fields has so far prevented their development,[52] but a tanker loaded oil from the Varandey field near the coast in October 1985.[53] Production is planned to commence in the Arkhangelsk district in the present plan period.[54]

The 'Peschanoozerskoye' oil and gas field on Kolguev Island is quite insignificant in terms of reserves, but very important for evaluating the surrounding shelf. Drilling on this island started in 1972. About ten productive zones have been found in triassic formations. Test oil production has reached 1095 barrels per day, and up to 320 000 cubic metres per day of natural gas has been produced.[55]

Further to the east, the Kara Sea—the offshore continuation of the West Siberian basin—is expected to contain significant petroliferous deposits. It is believed, however, that the chances of finding gas instead of oil are great since this offshore area can be regarded as an extension of the Yamal peninsula with its giant gas fields. The northernmost commercial gas find is reported to have been made on Byely Island to the east of Novaya Zemlya.[56] Soviet exploration activity is expected to move into the Kara Sea before 1990. Soviet exploration drilling on Svalbard, which has been going on for several years, also gives an indication of interest in the geology of the Barents Sea.

Other offshore areas In addition to the areas mentioned, offshore exploration is also taking place in the *Black Sea*. Three or four gas finds have been made. The Golitsyn field west of Crimea has been on stream since 1984. Two gas fields are in production in the shallow and ice-infested *Sea of Azov*, an arm of the Black Sea. There are expectations that oil will be discovered. The Soviets have been exploring in the *Baltic Sea* since 1984. Some small finds have been made off Kaliningrad. One field is on stream.

However, none of these areas has the potential to make them interesting in the context of a major supply alternative.

A brief comparison of the alternatives

The published parts of the new five-year plan are very vague on concrete exploration plans. It says that one is to 'speed up geological exploration after oil and gas on the territory of Western and Eastern Siberia, the Pricaspian Depression, in the European North, Central Asia and the Far East'. The Continental Shelf is mentioned in the context of more fundamental steps: one shall 'see to a well-planned technical re-equipment of the geological service, creation and installation of specialized, highly mechanized and automatized means for search and exploration for useful minerals, including for study and industrial utilization of natural resources on the Continental Shelf and in the World Ocean.'[57]

More immediate plans for the oil industry would 'continue the development of the oil industry in Western Siberia, Kazakhstan and in the north of the European part of the country; initiate industrial development of oil fields in deep horizons in the Pricaspian Depression; speed up the exploitation of oil and gas deposits on the continental shelf and create the necessary industrial-technological basis for this'.[58] Thus the plan does not announce any clear priorities and there is likely to be a struggle for investment capital and political attention.

In terms of supply potential, the areas can be divided into two groups: West Siberia and the Caspian Sea stand out as alternatives that can offer short-term results, but the potential in a longer perspective is limited. Eastern Siberia and the Barents Sea are much more uncertain, but the potential is possibly much larger. Kazakhstan falls somewhere in between. Again we see the problem of short-term gains versus long-term solutions. If there were no limitations to investment, the obvious thing to do now would be to secure output in the short-term simultaneously exploring to procure supplies for the longer term. To a certain extent this will be done, but some crucial decisions must be made to set economic and political priorities.

In the next chapter we will take a closer look at the Barents Sea. We use an investigation of activities in the area as one more indicator of the position of this region among the different production alternatives open to the Soviet Union.

Notes

1. Jonathan Stern, 'Western Forecasts of Soviet and East European Energy Over the Next two Decades (1980–2000)' in *Energy in Soviet Policy*, Joint Economic Committee, United States Congress, 1981.
2. Central Intelligence Agency I, *Prospects for Soviet Oil Production* and II, *Prospects for Soviet Oil Production: A Supplemental Analysis*, Washington, D.C., 1977. Admittedly the agency revised their estimates in 1979 and again later, but the two first reports had a much larger impact on scholarly discussion. For a complete history of the different CIA estimates and projections, see Marshall I. Goldman, *The Enigma of Soviet Petroleum: Half-empty or Half-full*, London, George Allen & Unwin, 1980, pp. 176–80.
3. See e.g. *Oil & Gas Journal*, 'Worldwide oil', 31 December 1984; E. N. Tiratsoo, *Oilfields of the World*, Beaconsfield, Scientific Press, 1984; Central Intelligence Agency, *USSR Energy Atlas*, Washington D.C., 1985.
4. Matthew Sagers, 'Natural Gas Liquids and the Soviet Gas Processing Industry', *CIR Staff Paper*, Washington D.C., Bureau of the Census, 1986.
5. *PlanEcon*, 'Soviet and East European Energy Databank', January 1986; and PlanEcon Report 14, 1986.
6. *PlanEcon*, 'Soviet and East European Databank', January 1986.
7. *Sotsialisticheskaya Industriya*, 9 March 1986.
8. Jan Vanous, 'Soviet Foreign Trade Performance in 1984' in *Centrally Planned Economies Current Analysis*, Wharton Econometrics, Washington D.C., 16 April 1985.
9. Ibid.
10. Central Intelligence Agency, *Handbook of Economic Statistics 1985*, Washington D.C., 1985.
11. *PlanEcon Report* 12, 1986.
12. *PlanEcon Report* 17–18, 1986.
13. Calculations based on figures from *PlanEcon*, 'Soviet and East European Databank', January 1986.
14. *PlanEcon Report* 7, 1986.
15. 'Long-Term Trends in Energy Consumption and Trade in the Soviet Union and Eastern Europe', *PlanEcon Report* 12–13, 1986.
16. Ed A. Hewett, *Energy, Economics, and Foreign Policy in the Soviet Union*, Brookings Institution, Washington D.C., 1984, p. 113.
17. *Osnovnye polozhenya energeticheskoy programmy SSSR na dlitelnuyu perspektivu*, Moscow, Politicheskaya Literatura, 1984.
18. L. A. Melentyev and A. A. Makarov *Energeticheski Kompleks SSSR*, Moscow, Ekonomika, 1983, pp. 167–8. According to Ed

Hewett the apparent aim of the authorities is to switch two power stations a year from oil to gas, freeing 250 000 barrels per day (2.5 million tons a year). (Lecture at the Fridtjof Nansen Institute, Oslo, March 1985.)

19. Matthew J. Sagers and Albina Tretyakova, 'Constraints in Gas for Oil Substitution in the USSR: The Oil Refining Industry and Gas Storage', *Soviet Economy*, 1, 1986.
20. *Handbook of Economic Statistics 1985*, op. cit.
21. *PlanEcon Report*, 19–20, 1986.
22. Soviet statistics as compiled by Hewett, op. cit., 1984.
23. The plan was first published in a draft version, announcing the aggregate investment figures, in November 1985 (see for example *Sotsialisticheskaya Industriya*, 9 November 1985). A revised version, after the 27th Party Congress, was published in March (see e.g. *Sots. Ind.*, 9 March 1986). The final information on energy investments was given by the Chairman of the Council of Ministers, Nikolay Ryzhkov, in a speech at the session of the Supreme Soviet formally accepting the plan, *Pravda*, 19 June 1986.
24. *Ekonomicheskaya Gazeta*, 48, 1985.
25. It is very likely that the gas industry will continue its habit of overfulfilling its plan, so that final results in the 12th five-year plan will be somewhat above 32 per cent growth.
26. See Thane Gustafson, 'The Origins of the Soviet Oil Crisis', *Soviet Economy*, 2, 1985.
27. See for example M. Gorbachëv's speech in Western Siberia, *Sotsialisticheskaya Industria*, 9 September 1985: 'Not today it became clear that the days of the "golden fountains", or if one can put it that way, the days of easy oil are coming to an end, that it is necessary to change to enhanced recovery, go into more difficult regions, into fields with lower flow rates, to conquer more difficult deposits. Already then it was also necessary to understand everything more deeply, to look ahead and carry out a programme intended for the new stage of capturing deposits. But this was not done in time. Central administrative organs, Soviet and party organs in Tyumen showed slowness and in overcoming the difficulties chose the path of minimum resistance, by deciding to compensate their poor preparations by increasing the load on the big fields. For work in the new difficult conditions not only the oil workers showed themselves unprepared, but also builders, energy specialists and transport workers.'
28. For an excellent account of the problems in Western Siberia see Thane Gustafson, op. cit.
29. B. Sidorenko, 'Problemy Tyumenskoy nefti', *Ekonomicheskaya Gazeta*, 9, 1986.

30. See *International Petroleum Encyclopedia*, Vol. 18, Tulsa, PennWell Publishing Co., 1985, p. 236.
31. See for example A. A. Trofimuk and S. M. Nikolajev, *Sibirien: Geographische Gegensatze, Mineralressourcen und Probleme deren Ersschliessung*, Novosibirsk, 1982; see also 'Rezervy geologov Tyumeni', *Sotsialisticheskaya Industriya*, 4 September 1985. It is claimed that these reserves have not been explored due to lack of equipment for drilling between 4000 and 6000 metres.
32. A common understanding of 'giant oil field' is a field with more than 500 million barrels (68.5 mt) of ultimate recoverable resources. A 'supergiant' has more than 5 billion barrels (685 mt). Central Intelligence Agency, *USSR Energy Atlas*, Washington D.C., 1985.
33. 'S chego nachinayetsya promysel', *Sotsialisticheskaya Industriya*, 1 February, 1986.
34. A recent Soviet press report can be taken as an indication of this, 'Po narastayushchiy', *Sotsialisticheskaya Industriya*, 13 May 1986. In this article it is claimed that by May 1986 17 new fields in Western Siberia have been put on stream, four above the plan. But the reserves are poorly utilized, due to lack of power supply (for water injection) and infrastructure in general.
35. *Stroitelstvo Truboprovodov*, 6, 1986, p. 1.
36. Average figures from Albina Tretyakova and Meredith Heinemeier, 'Cost Estimates for the Soviet Oil Industry 1970 to 1990', *CIR Staff Paper*, Washington D.C., Bureau of the Census, 1986. Costs in marginal fields: personal communication from Albina Tretyakova.
37. See for example *Petroleum Economist*, May 1985. (Aganbegyan now holds an important position as economic adviser to Gorbachëv in Moscow. See also *Platt's Oilgram News*, 5 February 1986.
38. See for instance Arthur A. Meyerhoff, 'Soviet Petroleum: History, Technology, Geology, Reserves, Potential and Policy' in Robert G. Jensen, Theodore Shabad, and Arthur W. Wright (eds), *Soviet Natural Resources in the World Economy*, Chicago, University of Chicago Press, 1983.
39. *Soviet Geography*, April 1986.
40. K. I. Isangulov and R. A. Maksutov, 'Tekhnika dlya osvoenya kontinentalnogo shelfa', *Gazovaya Promyshlennost*, 6, 1985.
41. *Soviet Geography*, April 1986.
42. See *Oil & Gas Journal*, 19 May 1986. p. 44.
43. *OGJ*, 10 March 1986.
44. *Soviet Geography*, April 1986.
45. See for example *Platt's*, 31 March, 1986. For a more detailed discussion of the project see Jonathan Stern, *Natural Gas Trade in*

North America and Asia, London, Gower, 1985, pp 197–201; Peter Egyed, Western Participation in the Development of Siberian Energy Resources: Case Studies, Ottawa, Carleton University, 1983. An up-to-date account of the impact of broader economic relations on the project is given by Leslie Dienes, 'Soviet-Japanese Economic Relations', *Soviet Geography*, September 1985.

46. A. N. Trutnev, 'Sakhalinskoye Uskoreniye', *Gazovaya Promyshlennost*, 5, 1986.
47. An official from Sakhalinmorneftegazprom (the local production organization) recently complained in a newspaper article that they were compelled by the gas ministry to fulfil the production plan, even if the demand was less than production. Consequently, they had to waste the gas. I. Khizhko, 'Zaplanirovannyj pereraskhod', *Sots. Ind.*, 14 April 1986.
48. A. N. Trutnev, op. cit.
49. Jonathan Stern, op. cit. p. 200.
50. Chingiz G. Sattarov, Gosplan, 'Statement on the current state and prospects of the oil and gas development on the outer continental shelf of the USSR.' Paper presented at the Fridtjof Nansen Institute, Oslo, 14 June 1984.
51. Meyerhoff, op. cit, p. 336.
52. *OGJ*, 20 August 1984. Recent reports on discoveries in this area have turned out to contain nothing new, probably referring to older onshore finds. See for example *Financial Times*, 9 October 1985.
53. *Soviet Geography*, March 1986, p. 200.
54. Oil minister Dinkov at the 27th Party Congress, *Pravda*, 3 March 1986.
55. *International Petroleum Encyclopedia*, Vol. 18, Tulsa, PennWell Publishing Co., 1985, pp. 233–5.
56. See *Oil & Gas Journal*, 20 August 1984, pp. 72, 19 February 1983, p. 72.
57. *Sots. Ind.*, 9 March 1986.
58. Ibid.

2 Soviet oil exploration in the Barents Sea

This chapter presents a detailed description of Soviet exploration efforts to date in the Barents Sea, with particular focus on the build-up of equipment and the use of foreign machinery and services. On the basis of these analyses, we evaluate the status and future prospects of the Soviet offshore venture, comparing it with a parallel development on the northern Norwegian shelf.

Seismic surveys

Soviet geological exploration of the Barents Sea started as early as 1971–72, and since 1978 extensive surveys of industrial (multifold) standard have been carried out, mostly between Novaya Zemlya and the Kola Peninsula[1]. The activity has been gradually stepped up in recent years. According to one Soviet source, kilometrage was expected to double from 1981 to 1985, as against the previous five-year plan. Total coverage was estimated at 130 000 kilometres of regional and sub-regional prospecting profiles by mid-1984.[2] At that time, seismic activity in the southeastern Barents Sea remained insignificant, but it was indicated that 'these rather prospective oil and gas bearing shallow waters' would be given priority in the future[3].

Based on these surveys, Soviet scientists have established the main features of the geological structure of the southern and southeastern parts of the Barents Sea. A number of local structures have also been identified in the southwestern part, among them Murmanskaya (Murmansk High) and Severo-Kildinskaya. According to the same source, seventeen structures have been identified in the Barents Sea and two in the Kara Sea, and 7335 square kilometres had been prepared for exploratory drilling as of 1 January 1984. As the geological value of these data is dubious, the identification of these structures is probably more important for bureaucratic purposes, to indicate future priorities for exploration[4].

The physical coverage of seismic surveys in the Soviet Barents Sea is thus quite extensive, but it is highly uncertain as to what extent valuable and practical geological knowledge has been extracted. The Soviets seem to feel fairly self-sufficient about having mastered the equipment, parts of which have been imported from the West, but Western observers are very sceptical about Soviet ability to process the data they have acquired[5]. As the seabed in this area is particularly hard, seismic exploration requires an advanced computer processing programme to be effective. This problem has been solved in the Norwegian part of the Barents Sea, but there is good reason to doubt that the USSR has the necessary software and skilled manpower. So far, Soviet authorities have to our knowledge not been willing to let Western companies re-analyse their seismic data.

Exploratory drilling

The first drilling operations in the Soviet Barents Sea, beyond very shallow waters, were begun in May 1982 by the Finnish built drill ship *Valentin Shashin* on Murmansk High (all locations are indicated in figure 2.1[6]. As the Soviet crew had great difficulties in operating the drilling gear, the well penetrated only the upper part of the section. The riser broke during the operations inflicting damage on the equipment on board, and the ship subsequently had to go to Finland for repair.

In 1983, both *Shashin* and the new drillship *Victor Muravlenko* (see below for details on equipment) were operating in the Barents Sea. The well spudded at Murmansk High was completed by the *Muravlenko*. On 1 July 1983, the same ship was observed drilling a second well not far from the first. By 1 January 1984, drilling had reached a depth of 1900 metres, falling short of the planned 3300 metres. Shortly afterwards, in January 1984, the *Muravlenko* had to leave for repair as the riser had again broken. In early May 1983, a third well was spudded at 248 metres by the *Shashin*, practically on the demarcation line to the disputed area between Norway and the Soviet Union in the area called by the latter Severo-Kildinskaya.

In 1984, the *Muravlenko* restarted drilling on Murmansk High, spudding a new well. By the end of July, there were

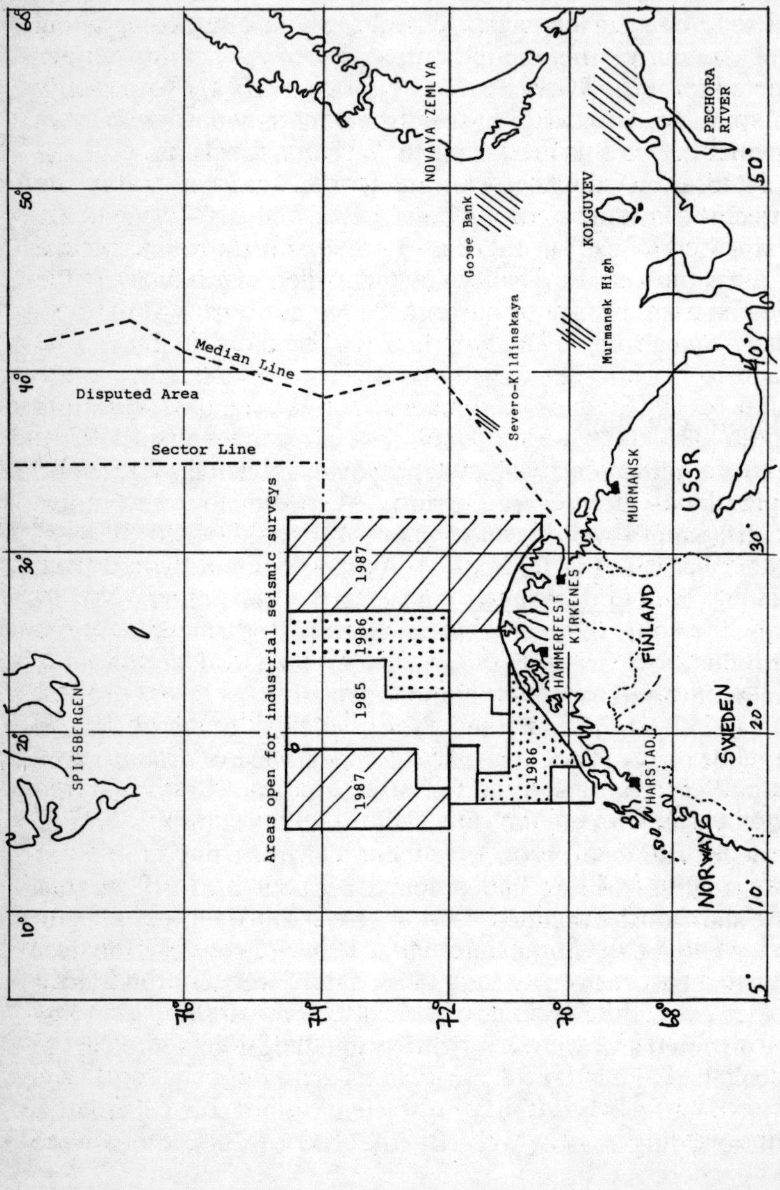

Figure 2.1 Exploration areas in the Barents Sea

Sources: Norwegian: Report to the Parliament no. 79, 1984–5, p. 41; Soviet: FNI estimates

rumours of an oil find, but these were never confirmed. Another two wells were spudded, probably by the *Muravlenko* in 1984, but the location is unknown. According to local officials in Murmansk, one of the vessels, probably the *Muravlenko*, was still drilling in the beginning of December. Operations from 1982 to 1984 had the character of an exercise aimed at giving Soviet operators experience in offshore exploration under Arctic conditions. By the end of 1984, six wells had apparently been spudded, but hardly more than one or two completed, a performance quite unacceptable by Western standards.

In 1985, exploration activity was stepped up considerably: the *Muravlenko* continued drilling on Murmansk High (close to the well spudded in 1983), and a new jack-up *Kolskaya* was placed near the mouth of the Pechora River. A new semi-submersible *Shelf-4* started up operations on a new structure, the Goose Bank to the north of Kolguev Island. The *Shashin* had to be brought to Finland once again for repair, probably early in the autumn of 1985 where it remained for servicing at a Rauma Reppola yard until the following spring, perhaps even longer. Despite this, exploration results improved noticeably in 1985 compared with the previous seasons. According to our information, five or six wells have been completed, i.e. the planned depths have been reached approximately, three of them during autumn 1985, indicating that Soviet operators are more able to manage the equipment they have acquired, in particular the less sophisticated semi-subs and jack-ups. The continuing problems with the *Shashin*, however, show that Soviet operators are still unable to handle the technolgy of a drillship without long delays, even after four years' experience. The drilling season now apparently runs from March to November, even if the jack-ups should be able to stand the ice also through the winter.

Few details of the location of these drilling operations are known, but, as indicated above, we have reason to believe that two or three wells have been completed on Murmansk High, one or two on the Goose Bank, one close to the disputed Norwegian/ Soviet area and one or two in shallow waters in the southeast (see figure 2.1). As to the findings of the exploratory drilling, even less is known. One Soviet source reveals that the first well on Murmansk High tapped middle triassic deposits and encountered two gas producing intervals[7]. It is also believed that a gas find was made in the well drilled by the *Kolskaya* near the outlet of the Pechora River. Both the seismic surveys and the location of the drilling operations indicate basically four prospective petroliferous areas in the Soviet part of the Barents Sea: the shallow waters in the southeast, the Goose Bank, Severo–Kildinskaya near the disputed area and Murmansk High.

Input of resources in offshore exploration

The first Arctic offshore drilling was probably made from the *Sevastopol*, a ship converted in the USSR to an ice-resistant fixed platform grounded near the mouth of the Pechora River.[8] It has not been in operation for the last two or three seasons, apparently because of ice problems. Part of the *Sevastopol* crew is reported to have been transferred to the *Viktor Muravlenko*. In addition, two fixed platforms seem to have operated near the mouth of the Pechora River up to late 1982, or early 1983.

Mobile drilling rigs

(For definition of rig types see figure 2.1)

Drill ships In 1979 the Soviet foreign trade organization Sudoimport ordered three drillships from the Finnish construction company Rauma Reppola[9], for a contract sum reportedly equal to FIM 900 million[10]. These ships are designed for drilling in Arctic waters at a depth of 300 metres, down to 6000 metres[11], and equipped with dynamic positioning systems from the Norwegian firm Kongsberg Våpenfabrikk.[12] The drillships are considered the world's most advanced for Arctic conditions.[13]

The first, the *Valentin Shashin*, was delivered in December 1981[14] and put into operation in the Barents Sea during 1982. According to Soviet sources, the second, the *Victor Muravlenko*, was ready for use in the Barents Sea in July–August 1982.[15] It is unclear whether the vessel managed to spud any wells during that season. It may have started drilling, but did not complete any wells. The third ship, the *Michail Mirchink*, was to be delivered in September 1982, and after tests to 'start probing the Arctic shelf'[16], it was put into operation in Vietnam[17]. No reports indicate drilling by the *Michail Mirchink* in the Barents Sea.

The *Valentin Shashin* and the *Victor Muravlenko* have been based in Murmansk since 1982, but their operations have been interrupted on several occasions for repair and maintenance in

Finland, as described above. Norwegian experts were brought in to assist the Soviet operators in handling the dynamic positioning equipment in 1983 and again in autumn 1985.

Semi-submersible platforms There have been conflicting reports on how many semi-submersible 'Shelf' rigs are being built at the Soviet yard in Vyborg—four, or less.[18] It now seems clear, however, that the 'Shelf' rigs numbered *1*, *2* and *3* have been built in Astrakhan at the Caspian Sea, that numbers *4* and *5* have been launched, and *6* is under completion in Vyborg. The Shelf exploration drilling rigs are of modified Dutch 'Pacesetter' design[19]; they should be able to operate in Arctic waters at 200 metres' depth, drilling down to 6000 metres.[20]

The *Shelf-4* underwent lengthy tests in the Baltic Sea at the end of 1984 and early 1985 before being towed to Kolguev Island to be launched[21]. It is now possibly operating off the coast of Vietnam. At the end of 1984 it was reported that work on *Shelf-6* was 'proceeding rapidly', but no announcement has been made as to where it will operate.[22] Construction of *Shelf-7* is also believed to have started. Long delays in testing the new rigs have been reported.[23] In the summer of 1985, negotiations between Rauma Reppola and the USSR on a cooperative arrangement to build semi-subs were reported: The Vyborg shipyard was to build and deliver the superstructure to Rauma Reppola's yard on the west coast of Finland for completion.[24] Apparently, no contract has been signed yet.

Jack-up rigs The first of two Arctic class jack-up drilling rigs ordered by Sudoimport and produced by Rauma Reppola was named the *Kolskaya* in January–February 1985.[25] It probably arrived in the Barents Sea in July 1985 and is now operating near the mouth of the Pechora River.[26] The second jack-up, the *Sakhalinskaya*, has—not surprisingly—been put into operation off Sakhalin. The plan apparently is to use three jack-ups in the shallow waters (down to 100 metres) in the southeastern Barents Sea.

On 22 March 1985, Rauma Reppola signed a new agreement with Sudoimport for delivery of six legs, eighteen racks and other parts for another two Arctic-class jack-ups to be constructed at the Soviet yard in Vyborg. Each leg is to be about 140 metres

Table 2.1 Soviet mobile offshore drilling rigs

Area	Until summer 1986	Configuration summer 1986	1986–1990 Expected future supplies	1990
Barents Sea[1]	1982: *Valentin Shashin* ds 1982: *Viktor Muravlenko* ds 1985: *Kolskaya* ju 1985: *Shelf-4* ss	2 drillships 1 jack-up 1 semi-sub (4)	86–7: 1 *Shelf* ss 87–8: 2 jus 87–90: 2–5 additional rigs	9–12
Sakhalin[2]	1983: *Mikhail Mirchink* ds[1,14] 1980: *Okha* ju[3] 1984: *Ekhabi* ju[14] 1985: *Sakhalinskaya* ju[1]	1 drillship 3 jack-ups (4)	86: *Shelf* 6ss[16] 87–90: 1–2 additional rigs	6–7
Caspian Sea[4]	1981: *Kaspmorneft'* ss[5] 1981: *Shelf-1* ss[5] 1983: *Shelf-2* ss 1986: *Shelf-3* ss[6] –1974: *Apsheron, Azerbaydzhan* and *Khazar* ju's[5] 1975: *Baku* ju[5] 1978: *60 let Oktyabrya* ju[5] 1981: *60 let Azerbaydzhana* ju[5] 1985: *40 let Pobedy* ju[7]	4 semi-subs 7 jack-ups (11)	86–90: 5–10 additional rigs[11]	16–21
Black and Azov Seas	1981: *Sivash* ju[5,8]	1 jack-up[15] (1)	86–90: 0–3 additional rigs (1–2 Romanian)	6–9
Baltic Sea[10,5]		1 jack-up	86–90: 0–2 additional rigs	1–3

TOTAL	*21 units*			35–52
	(13 jack-ups,	Soviet domestic prod.	86–90: 16	
	5 semi-subs	Romanian jus[12]	1–2	
	3 drillships)	Brasilian jus[13]	4	
			21–2	

Foreign rigs: Dutch (1) *Khazar*
Japanese (1) *Okha*
Singapore (1) *Ekhabi*
Finnish (6) *Kaspmorneft, V. Shashin, V. Muravlenko, M. Mirchink, Kolskaya, Sakhalinskaya*

ju = *Jack-up.* A floating drilling platform with legs. The platform floats while being moved. But the legs can be lowered and the deck jacked up to allow the platform to stand on the ocean floor while drilling.

ss = *Semi-submersible.* A drilling platform consisting of a body under water for stabilization and at deck with drilling rig. The deck is attached to the body by pillars minimizing the diameter at the water-line, thus reducing the movements of the vessel.

ds = *Drillship.* A ship with a drilling rig, particularly for drilling in deep waters. To stabilize the ship it has to be equipped with a dynamic positioning system.

1 See description in this chapter.
2 A. N. Trutnev: 'Sakhalinskoye Uskoreniye', *Gazovaya promyshlennost*, 5, 1986, p. 7.
3 *Oil & Gas Journal* (*OGJ*), 12 October 1981, p. 170. The *Okha* platform was built in Japan.
4 *OGJ*, 19 May 1986.
5 *OGJ*, 12 October 1981, p. 16. The Dutch *Khazar* is the prototype for the following *Kaspii* series of Soviet built jack-ups. The *60 let Azerbaydzhana* is an improved version, which can drill 6,500 metres in 70 metres of water.
6 *Ekonomicheskaya Gazeta*, 14, 1986, p. 1; according to several sources the *Shelf-3* was completed long ago, but this recent article proves the opposite.
7 *Soviet Geography*, April 1985, p. 293.
8 *Sotsialisticheskaya industriya*, 14 May 1983.
9 *Gazovaya promyshlennost*, 12, 1985, p. 8.
10 *International Petroleum Encyclopedia*, vol. 18, 1985, p. 232.
11 *OGJ*, 19 May 1986; Azerbadzhany local officials ask to receive about 20–5 more rigs.
12 *Petroleum Economist* (*PE*), April 1984, p. 146.
13 *Norges handels og sjøfartstidende*, 7 March 1985. However, recent reports indicate that this project is cancelled.
14 The rigs have also been operating off Vietnam. *Petroleum News*, December 1985.
15 There is uncertainty concerning the characteristics of other installations operating in the Black Sea. The *Golitsyno* series consisting of 5 rigs has been referred to both as fixed platforms and as semi-subs; see for example *Gazovaya Promyshlennost*, 12, 1985. In any case, the waters in these areas are very shallow and the rigs probably not very sophisticated; thus we exclude them from this list of 'real' mobile drilling rigs.
16 *Sots. Ind.* 14 September, 1986.

long, platform weight 14 500 tons. Drilling depth capacity has not been disclosed.

The RR–Sudoimport contract comprises about one-third of the total construction costs, which are estimated to be FIM 900 million (approximately $170 million). Deliveries are set for 1986–87.[27]

The figures above show that the Soviet fleet of mobile drilling rigs has grown rapidly from only two units in 1975 to seventeen in 1984 and twenty-one at present (see table 2.1). It is also worth noting that a constantly rising share of the new rigs are built in the USSR, in Vyborg and Astrakhan.

Other equipment

Geological research In 1978 the USSR purchased relatively modern equipment for seismic survey in the Barents Sea from a leading French geophysical company and this has been in active use ever since.[28] Two vessels (5300 tons) designed for geological, oceanographic and related research are to be delivered in 1986–7 from the Finnish Hollming Company to the Soviet gas ministry, in addition to the three 2600 ton research vessels already on order at the same yard. They will operate on depths up to 300 metres and be able to drill in the seabed down to 200 metres. The vessels will be equipped with dynamic positioning systems.

Supply vessels The Ullstein shipyard on the western coast of Norway has delivered seven special supply vessels, named UT 700, to the USSR since 1976.[29] At least two are known to be based in Murmansk and operating in the Barents Sea. It has been reported that about twenty supply vessels are currently under construction at the Gdansk yard in Poland for Soviet use.[30]

Catamarans The first of four gigantic catamaran vessels was delivered from the Finnish construction company Wärtsila in September 1984 and launched in the Caspian Sea. The vessels are designed for offshore construction works. The last two vessels are to be completed at Soviet shipyards; it is not known if any will operate in the Barents Sea.

A heavy-lift vessel with capacity up to 20 000 tons has been ordered for delivery in late 1987 by the USSR at Wärtsila's Turku shipyard to transport semi-subs and jack-ups. This will facilitate communication between the exploration areas.[31]

Offshore base in Murmansk

Until now a part of the northern area of Murmansk harbour on the Eastern side of the Kola Fjord has been used as a temporary service base for Barents Sea offshore activity. Construction of a new base on the western side *vis-à-vis* the town was reported to be well under way in the middle of 1984.[32] This base is intended to have three functions: supply/service, repair and construction of offshore rigs. In connection with the base, a settlement for several thousand inhabitants is planned. There are also plans for a new bridge crossing the Kola Fiord to the north of the old one which will make communication easier between the town and the base area.

Manpower

At the outset of operations in the Barents Sea, the Finnish built drillships were manned by experienced oil workers from Azerbaydzhan; this was not very successful and the vessels are now operated by Russians. Technical problems do, however, still exist. The dynamic positioning system did not function without assistance from the producer during the first and second seasons. The drilling record of the *Shashin* indicates that some of these problems remain. To improve qualifications, a school for oil-workers has been established in Murmansk. It is planned to have some 1200 students and will also educate workers for sub-sea activities.[33] (see appendix 1) Some workers on the drillships are graduates of the Gubkin Institute for Petrochemistry and Gas Industry in Moscow.

Organization

In 1978 the gas ministry took over responsibility for the Soviet off-shore industry from the ministry of geology—apparently in an attempt to speed up exploration activities. The gas ministry (Mingazprom) is considered one of the most dynamic and successful organizations in the central bureaucracy having over-fulfilled its production plans every year since the mid-1970s. In addition to the gas industry—which is the largest in the world with an output of 643 billion cubic metres in 1985—the ministry produces some 20 million tons of oil annually.[34] Its investments in gas extraction was estimated at 4.2 billion roubles in 1985. Mingazprom has established a special directorate within its central appar-

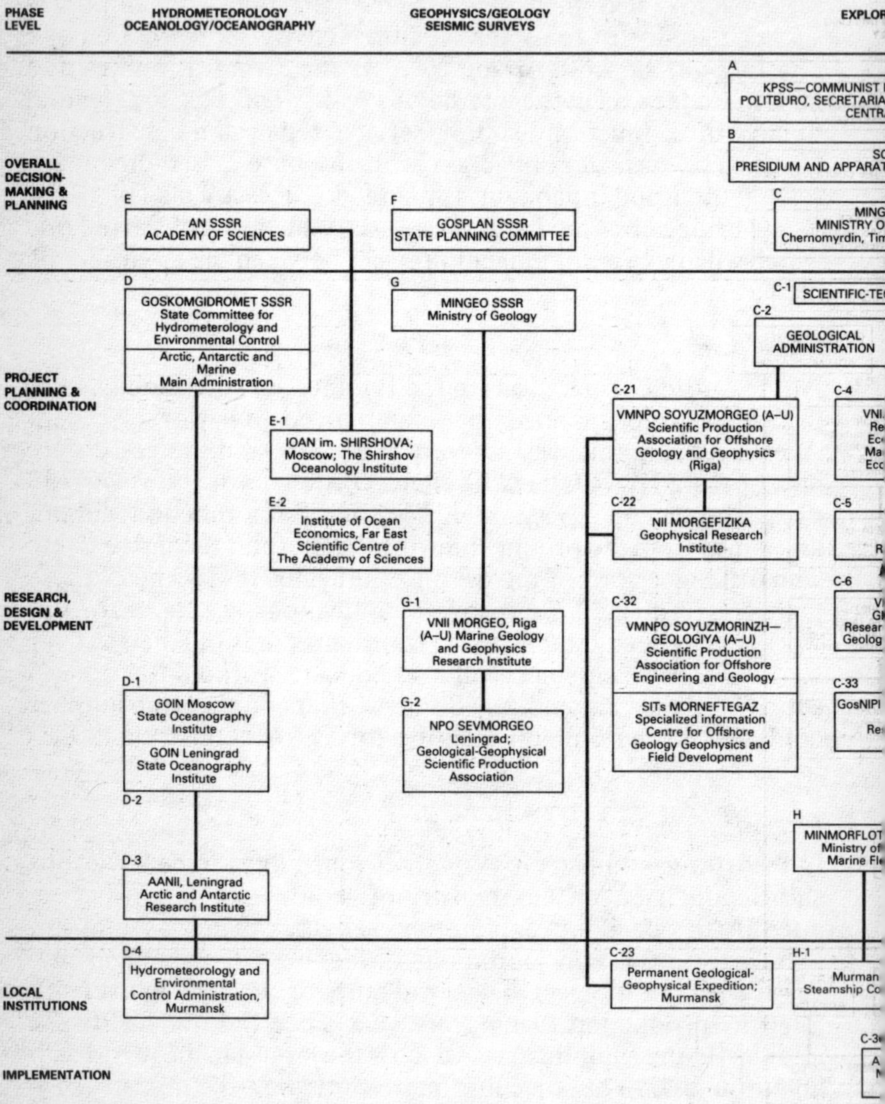

Figure 2.2 The Barents Sea Offshore Organization

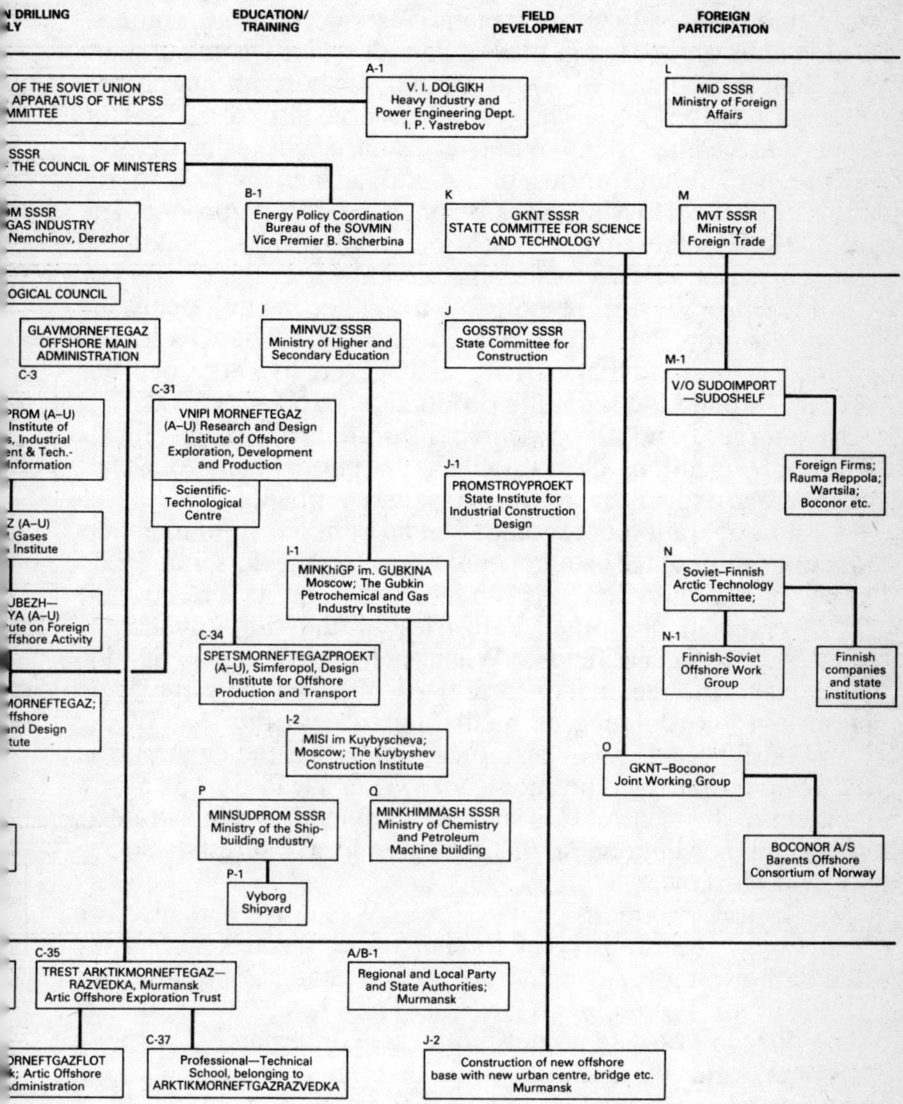

DRILLING
LY
EDUCATION/ TRAINING
FIELD DEVELOPMENT
FOREIGN PARTICIPATION
OF THE SOVIET UNION
APPARATUS OF THE KPSS
MMITTEE
A-1
V. I. DOLGIKH Heavy Industry and Power Engineering Dept. I. P. Yastrebov
L
MID SSSR Ministry of Foreign Affairs
SSSR
THE COUNCIL OF MINISTERS
M SSSR
GAS INDUSTRY
Nemchinov, Derezhor
B-1
Energy Policy Coordination Bureau of the SOVMIN Vice Premier B. Shcherbina
K
GKNT SSSR STATE COMMITTEE FOR SCIENCE AND TECHNOLOGY
M
MVT SSSR Ministry of Foreign Trade
OGICAL COUNCIL
GLAVMORNEFTEGAZ OFFSHORE MAIN ADMINISTRATION
I
MINVUZ SSSR Ministry of Higher and Secondary Education
J
GOSSTROY SSSR State Committee for Construction
M-1
V/O SUDOIMPORT —SUDOSHELF
C-3
PROM (A–U)
Institute of
s, Industrial
ent & Tech.-
Information
C-31
VNIPI MORNEFTEGAZ (A–U) Research and Design Institute of Offshore Exploration, Development and Production
Scientific-Technological Centre
J-1
PROMSTROYPROEKT State Institute for Industrial Construction Design
Foreign Firms; Rauma Reppola; Wartsila; Boconor etc.
Z (A–U)
Gases
Institute
I-1
MINKhiGP im. GUBKINA Moscow; The Gubkin Petrochemical and Gas Industry Institute
N
Soviet–Finnish Arctic Technology Committee;
UBEZH—
YA (A–U)
ute on Foreign
ffshore Activity
C-34
SPETSMORNEFTEGAZPROEKT (A–U), Simferopol, Design Institute for Offshore Production and Transport
N-1
Finnish-Soviet Offshore Work Group
Finnish companies and state institutions
MORNEFTEGAZ;
ffshore
nd Design
tute
I-2
MISI im Kuybyscheva; Moscow; The Kuybyshev Construction Institute
O
GKNT–Boconor Joint Working Group
P
MINSUDPROM SSSR Ministry of the Shipbuilding Industry
Q
MINKHIMMASH SSSR Ministry of Chemistry and Petroleum Machine building
BOCONOR A/S Barents Offshore Consortium of Norway
P-1
Vyborg Shipyard
C-35
TREST ARKTIKMORNEFTEGAZ—RAZVEDKA, Murmansk Artic Offshore Exploration Trust
A/B-1
Regional and Local Party and State Authorities; Murmansk
ORNEFTGAZFLOT
k; Artic Offshore
dministration
C-37
Professional—Technical School, belonging to ARKTIKMORNEFTGAZRAZVEDKA
J-2
Construction of new offshore base with new urban centre, bridge etc. Murmansk

atus, Glavmorenftegaz, which is responsible for all offshore activities. The regional organization in Murmansk, Arktikmorneftegazrazvedka, is the operational headquarters for the exploration venture in the Barents Sea (see figure 2.2). This organization, which is headed by O.O. Sheremeta, is estimated to have some 3–5000 employees based in the Murmansk region—some of them with long experience of Arctic drilling onshore in Siberia and other parts of the USSR. This organization has also carried out the drilling on Kolguev Island. We have reason to believe that it has set up its own sub-unit upravlenie, Arktikmorneftegazflot, to administer the fleet in Murmansk belonging to Mingazprom.[35] The Ministry of the Merchant Fleet (Minmorflot) provides additional transportation services to the offshore industry.

Mingazprom is also responsible for collecting and analyzing seismic survey data from the Barents Sea. The Geological Administration of the ministry is believed to carry out this activity through the scientific production association, VMNPO Soyuzmorgeo, and the research institute, NII Morgeofizika. These organizations have established a permanent geological/geophysical expedition in Murmansk (see Appendix 1, 'C-23').

How cooperation between these and various other ministries functions in practice is largely unknown to outsiders, but we do know that Mingazprom has run into severe difficulties because of resistance from other industries in the development of Western Siberian gas fields[36]. When such bureaucratic infighting takes place, the regional or local party organization can play an important coordinating or mediating role. In the case of the Murmansk region, it is worth mentioning that the chairman of the regional party committee (*obkom*), V.N. Ptitsyn, is also a member of the Central Committee. Combined with the strategic importance of the region, this may give local interests a strong voice in Moscow.[37]

As to foreign trade, contacts with Western companies are handled by the ministry of foreign trade, through its agency Sudoimport, the gas ministry and the State Committee for Science and Technology (GKNT). They have set up special committees to deal separately with the Norwegian consortium, Boconor, and the Finnish companies involved. How these contacts are coordinated among these bodies, a multitude of research institutions belonging to several different organizations[38] and Gosplan which oversees all large-scale investment plans, is not known in any detail. It is uncertain whether co-ordination is taken care of by one particular organizational unit (possibly a joint committee or working group) or through informal channels. At any rate, there are obviously close contacts mainly between Mingazprom, Gosplan, and the foreign trade ministry in dealing with Western companies interested in the Soviet offshore industry.

Soviet offshore activity: a preliminary evaluation

Soviet exploration in the Barents Sea can be evaluated according to various criteria: it can be compared with the performance of Soviet industry in general, with the record of the petroleum sector, or perhaps more appropriately with the gas industry specifically, since it is organizationally responsible for this activity. (We have analysed the links between the energy industries in general and the Barents Sea offshore venture in Chapter 1.) Since our purpose here is to evaluate the make-up, technical record and overall effectiveness of the exploration described above, we will use the corresponding activity on the northern Norwegian shelf in the Western part of the Barents Sea, the so-called Tromsøflaket, as a yardstick for comparison. Since the international offshore industry operating in Norwegian waters is considered among the world's most advanced, having access to the latest technological innovations in the petroleum industry, such a comparison will focus on both the achievements and the shortcomings of the Soviet offshore industry—and its future needs.

The Norwegian northern shelf: a brief review of recent developments

The natural environment for offshore exploration in the Norwegian northern waters has many similarities with the North Sea. Water depths vary from 100–300 metres and climatic conditions with regard to wind, waves and currents are not radically different. The major difference is the low temperatures in the north; these lead to ice problems on the platforms, frequent heavy fog creating new security and supply difficulties and so-called polar low pressures. These are not unlike tropical typhoons and therefore pose a new challenge to the security of offshore installations and vessels. Drifting ice and icebergs have so far not been an obstacle to offshore exploration, but as activity moves farther north and east, this problem also has to be taken into account (see map of ice extension in figures 3.1 and 3.2)[39].

Exploration drilling in the first concessions off northern Norway started in 1980 and at the same time seismic exploration was stepped up rapidly. The number of wells drilled at Tromsøflaket are as follows:

1980:	2
1981:	3
1982:	4
1983:	6
1984:	7
1985:	5
1986:	4
Total:	31

For 1987, the number is expected to be five to seven. Average depth is around 3000 metres. Only one well goes down to 5000 metres. It normally takes 100 days to complete the 'average type'; the latter took six months. The costs of these operations are as a rule of a thumb approximately NOK 1 million per day (NOK 7 equals roughly $1), which brings the total costs of drilling operations alone to around NOK 3 billion. Each drilling unit has been calculated as able to complete two to three wells per year. As year-round drilling has been introduced recently, this figure will naturally increase.

Seismic surveys today amount to 153 000 kilometres in the Norwegian Barents Sea[40], and will be stepped up considerably in the years to come. The Norwegian Government has decided to open up vast new areas between 24 degrees and 32 degrees East for seismic exploration in 1986 and 1987. New concessions north and northeast of Tromsøflaket (Bear Island South and Finnmark West), will be awarded in the eleventh round in early 1987 (see figure 2.1). The results of the drilling in those concessions will be decisive for future exploration activity in the Norwegian Barents Sea. Furthermore, the companies have suggested that this part of the shelf be opened for 'strategic drilling' in the most promising areas, and the government's response seems to be positive.

This development is connected to a renewed interest in petroleum exploration on Svalbard, where 350 kilometres seismic was shot in 1986. A number of Norwegian oil companies (notably Norsk Hydro) and international ones (including BP), are now planning and carrying out research programmes on Svalbard of considerable magnitude for the islands[41]. The Norwegian Petroleum Directorate (NPD) has shot seismic in the waters between Tromsøflaket and Svalbard and eastwards to the North Cape basin, with reasonably positive results. We take all these factors to indicate a constantly growing interest on the part of the oil industry in the Barents Sea as a whole. It is worth noting that in addition to Norwegian companies, both Shell and Esso are operators on Tromsøflaket, while Elf, Mobil, Tenneco, Conoco, Amerada and Fina have minority shares in the concessions awarded in this area in the ninth round. Whether this interest in oil exploration off northern Norway and on Svalbard will survive the present price fluctuations in the world oil market remains to be seen.

All these activities must be seen as a natural extension of the offshore industry operating in the North Sea. It has been estimated that at least 90 per cent of the input into northern shelf operations originates from this source. Local and regional activities are largely confined to the recently established bases at Harstad, Andøya and Hammerfest (see figure 2.3). They have sufficient capacity to service all the planned exploration in the Norwegian sector in the North. So, from a purely commercial point of view, there is no need for new bases in northern Norway. However, this is a sensitive political issue as there is strong regional pressure for more diversified industrial activity in the North. If the Soviet offshore industry in the Barents Sea should demand base services from the West, this will obviously stimulate efforts to establish a large offshore base in Kirkenes—for which various preparations have already been made at the local level.[42]

A number of gas finds have been made at Tromsøflaket, but so far, very little oil. According to NPD estimates, gas reserves which can be produced with present technology may amount to 250 million tons oil equivalent[43]. No figures for oil reserves have been given. Nobody is at present willing to estimate how large an oil field on the Northern shelf would have to be to be considered commercially interesting, given today's volatile oil prices. It may

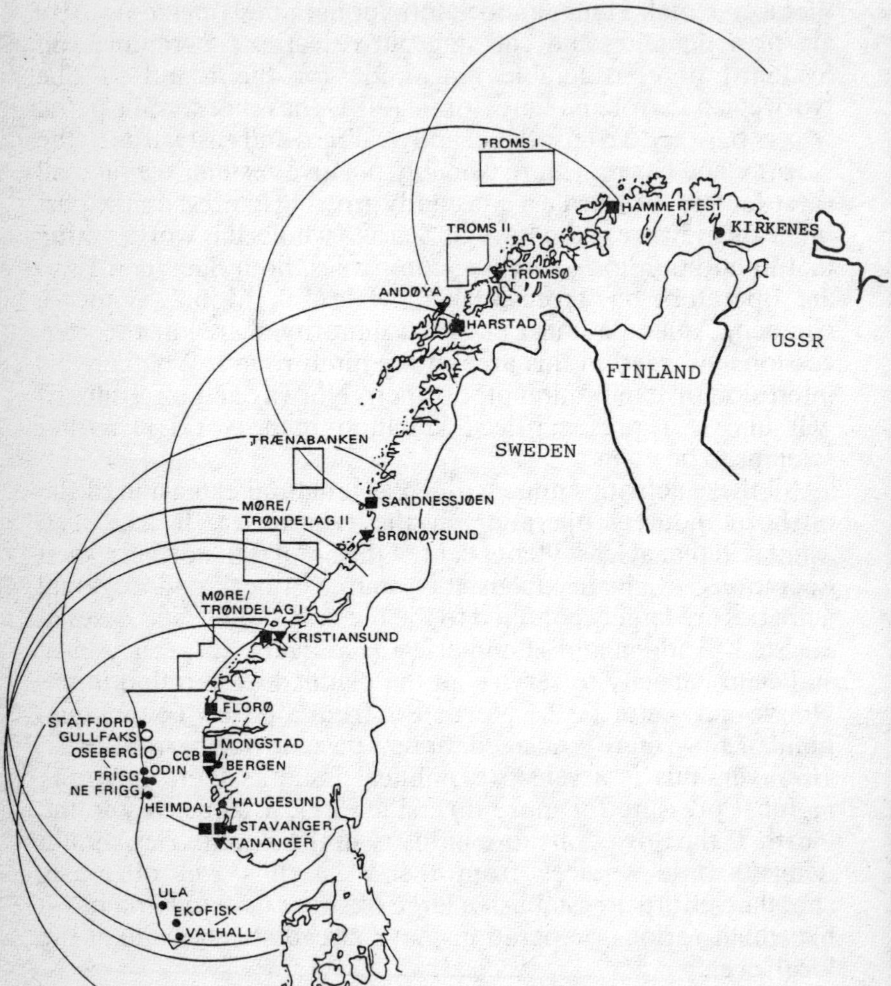

Figure 2.3 Norwegian supply bases
Source: Report to the Parliament no. 79, 1984–5, p. 22

be reasonable to presume that such a find would have to be at least as large as the Snorre field off the west coast, probably considerably larger. (Total oil reserves for Snorre are presently estimated at some 100 million tons. Despite this, development of the field has been postponed because of the fall in oil prices.)

Soviet and Norwegian exploration: a comparison

The differences between Soviet and Norwegian petroleum activity in the Barents Sea are striking. The Norwegian exploration has proceeded rapidly northwards from its base in the North Sea, where the entire industrial infrastructure is already in place after twenty years' experience. Companies have been eager to move into a new oil province with wholehearted support from the government—not to speak of the regional authorities who have been looking eagerly for a share of the oil bonanza so far reserved for the central part of the country. The major problem is the high cost of exploration. Technical adjustment to the northern environment seems to have advanced smoothly in spite of a few squabbles over 'winterization' standards of some drilling units. The effectiveness of the drilling operations (number of wells drilled per unit per season) and their being extended to a year-round season are the best indicators of this development.

The Soviet offshore venture has, in contrast, run into major technical problems. The drillships used the first two or three seasons mainly as an exercise to gain experience. Even in 1985, the fourth season, results were very modest by Western standards: four drilling units completed four or five wells, most of which were already spudded. This indicates that the effectiveness of the drilling equipment in use on the Soviet side is roughly one-third of the Norwegian average. According to Western standards, each drilling unit should be able to complete three wells each year, which implies that the USSR should have finished twenty to twenty-five wells by now instead of five or six. With year-round drilling in Norway, and no reliable reports on similar developments in the Soviet Union, this disparity will become even wider. This sluggishness on the Soviet side is largely due to inexperience with offshore oil under these conditions. The Soviet Union has no offshore industry of its own comparable to that in the West, so it lacks the technical, organisational and physical infrastructure for such a venture.[44] The USSR has to train or re-train personnel specifically for these tasks, to establish or reconstruct production capacity onshore and build up the necessary supply, maintenance and repair services—alone or in cooperation with Western partners.

The Soviet performance with regard to seismic exploration is very difficult to evaluate, owing to lack of accurate information. Coverage is, as mentioned above, quite impressive and compares very well with the Norwegian record, but the quality is largely unknown. Western experts estimate, however, that it is inferior to their standards.

When it comes to construction and deployment of drilling units, over the last few years the Soviet gas industry has built up a fleet which in numbers is fully adequate by Western standards: two drillships and two or three floating platforms, probably increasing to four or five[45], should be able to carry out the necessary exploration in the Barents Sea within reasonable time. Western experts also point out that the location of the wells seem well founded on present geological knowledge.[46]

So, with regard to planning and input of equipment, the Soviet offshore venture has the character of a deliberate and well calculated effort. Such underlying ambitions are, however, not matched by performance in the field as demonstrated by the dismal drilling record described above.

We have little accurate information concerning the exact reasons for these problems. We do know that there were severe problems during the first drilling seasons with handling the dynamic positioning equipment mainly because the personnel was not trained for this task. In 1984 the journal of the gas ministry—*Gazovaya Promyshlennost*—reported the main findings of an audit report of Arktikmorneftegazrazvedka which included heavy criticism of the management: the enterprise failed to fulfil its plans in 1983 because of 'low technological, financial and work discipline. The drilling results were only 70 per cent of the plan and because of low work productivity the principal investments were exploited by no more than 46 per cent'. The report even hinted at major accidents (*avarii*) causing significant material losses, without giving any details. This probably refers to the misfortune of the drillship operations in 1982–3. The officials responsible were punished—while the head of the organization, O. O. Sheremeta, received only a critical remark on his record—'in view of his short period of service'.[47]. We also have good indications of problems with coordination of supply and services, only too well known in Soviet industry. For example, when one of the Norwegian-built supply vessels was allowed to use a Norwegian port, the skipper is reported to have remarked: 'This cuts my down-time by 50 per cent'. When the modern fishing-fleet in Murmansk is stranded for long periods for lack of diesel fuel—a fact exposed on Soviet TV during the winter 1985–6—similar difficulties might also apply to the offshore vessels. Such problems will have an even more damaging effect in the offshore industry than elsewhere, as the management of numerous interlinked activities off and onshore is the key to effective operations. Such management also requires sophisticated computer equipment and specially trained manpower, of which there is certainly no surplus in the Soviet energy complex. In addition, as the Soviet petroleum industry is largely sealed off from the rest of the world, it does not automatically benefit from the continuous technological improvements taking place outside its borders.

Conclusions

This review of the Soviet offshore venture in the Barents Sea leaves us with the following conclusions based on the information available at present:

- The Soviet gas industry made a deliberate choice in the late 1970s to start petroleum exploration in the Barents Sea. For this purpose, the gas ministry ordered most of the necessary equipment to be built in Finland and with considerable input of Western technology purchased in hard currency. The construction works have been gradually transferred to the Soviet Union itself, taking advantage of existing productive capacity.
- The gas ministry has apparently used operations so far to learn how to handle new equipment and technology—in other words, transferring in a similar way the necessary skills from the West.
- The efficiency of these efforts has been very modest, indeed dismal by Western standards. This seems primarily due to inexperience and organizational/coordination problems which cause long delays. The leadership of Mingazprom is now taking the problem of inefficiency very seriously. The new gas minister, V. S. Chernomyrdin, has made it clear to his

staff that it is essential 'to increase significantly the efficiency of the work in offshore exploration and development'.[48] He has also underlined that 'it is imperative to make a thorough analysis of the work of the (offshore) fleet and to define measures needed to improve its use and diminish its costs, to secure a sharp improvement in the indicators for drilling'.[49] This position is also clearly reflected in the guidelines for the offshore industry published as part of the twelfth five-year plan (1986–1990) which reiterates the need for 'significant improvement of the technical-economic indicators for the development of geological exploration activity and oil production in the Glavmorneftegaz enterprises'.[50]

- There is reason to believe that the leadership of the gas industry has realized that they have run into severe problems, probably far more serious than they originally expected in the Barents Sea. They may have underestimated the specific difficulties facing oil exploration in this area. They may not have understood that it requires a large maritime, technological back-up environment and that it takes years to develop the necessary know-how. (After all, it took the Norwegian oil industry, with all its cooperation with the international petroleum industry over ten years to gain enough expertise to operate a major field.) Our findings indicate that the Soviet gas ministry sees the Barents Sea as an interesting option in the long term. Mingazprom is apparently taking the time it needs to build up an offshore industry in general and in the Barents Sea in particular. The ministry does not appear to be under any visible pressure to speed up these activities to a higher level. It is hard to find any strong organizational force, or coalition of forces, in the Soviet energy system interested in a radical increase in Barents Sea exploration and willing to commit the necessary resources for such a purpose. The uncertain resource potential of the area can easily turn this lack of momentum into a vicious circle: the uncertainty discourages input of resources - in particular lasting commitments like infrastructure—which hampers the exploration, in turn reducing the likelihood of finding a major field. It should be underlined in this connection that the USSR is searching for oil not gas in the Barents Sea, while the chances of finding gas seem to be higher. A combination of oil and gas will be attractive only if the oil reserves are large enough to warrant development of a field. (Gas from such a hypothetical find of condensate could be used locally in the Murmansk region or piped to the Finnish network which is already supplied with Soviet gas.)

An indication of the attitude of the gas industry to offshore oil and the Barents Sea option can be found in its purchases of foreign equipment. Appendix 2 presents a list of such orders in as much detail as possible, in view of the information available to us. In our judgement the following inferences can be drawn from this account: the Soviet gas industry has spent limited amounts of hard currency since 1979 on offshore related equipment. (It is impossible to give any precise estimate of the numbers involved.) These purchases have been heavily concentrated in the Finnish shipyard and equipment industries, and in hardware, such as vessels and drilling units. Even if Finnish equipment encompasses components from other Western companies, imports from Finland reduce Soviet hard currency expenditure. The scale of these orders however, indicates high priority on the Soviet side, since access to the Finnish equipment market is a scarce good in Soviet industry. Much of the equipment can serve different functions for economic development of Arctic areas, apart from offshore oil exploration. This may indicate that the gas industry is reluctant to make any clear-cut commitment to one particular option in its use of hard currency.

Future prospects

Table 2.1 indicates that the build-up of the Soviet fleet of mobile drilling rigs is planned to continue from 1986 to 1990, roughly doubling the number of units and simultaneously increasing the relative share constructed in the USSR. The Barents Sea seems to be getting its due share of the additions to the fleet (9–12 out of 35–52 units)[51]. These figures are in themselves an indication of a lasting interest in the offshore option in general and in the Barents Sea in particular. However, the shortcomings of the Soviet offshore venture so far are not due to insufficient drilling equipment. As shown above, the problem has been the quality and efficiency of operations, not the quantity of operating units. As the USSR increases its efforts—the figures above indicate a doubling of resource input by 1990—this disparity between quantitative growth and practical performance may become more obvious and acute. If this is correct, the USSR will be faced with some very hard choices in its offshore policy. For the purpose of this discussion, we can pinpoint three possibilities:

1. The Soviet gas industry can choose to go ahead with its planned build-up and run the risk of considerable, perhaps enormous waste of domestic resources for lack of services, infrastructure and effective management. If a major oil discovery is made before 1990, this will mean the same dilemma again, but with much higher stakes, since production requires several times the resources of exploration.
2. The Soviet authorities—within or above Mingazprom—may decide to save domestic resources for other purposes, thereby postponing the offshore and Barents Sea option into the next century. In this way they can take the time to develop the necessary skills and experience with fewer resources at stake. However, such a policy will reduce the likelihood of major progress in the Barents Sea to near zero.
3. The Soviet gas industry can make a deliberate choice in favour of large scale foreign participation in Barents Sea oil exploration and possible development. The preconditions and consequences of such a policy are discussed in Chapter 4.

Notes

1. Egil Bergsager (then deputy director of the Norwegian Petroleum Directorate) in a statement to a seminar organized by the Norwegian Petroleum Society, 21 October 1985.
2. Statement made by Chingiz G. Sattarov, chief expert, Gosplan, at the Fridtjof Nansen Institute, 14 June 1984. In comparison, seismic surveys in Norway amount to 100 000 kilometres per year on the whole shelf.
3. Sattarov, op. cit. The USSR may have encountered difficulties in the seismic exploration because of the shallowness.
4. Each structure is on average 450 square kilometres, compared with for example the Troll field of 750.
5. This is underlined by Robert W. Campbell in 'Technology Transfer in the Soviet Energy Sector' in Bruce Parott (ed.) *Trade, Technology and Soviet-American Relations*, Bloomington, Indiana University Press, 1985, pp. 156–7.
6. TASS press release, 30 July 1982.
7. Sattarov, op. cit.
8. *Oil and Gas Journal*, 26 April 1982, p. 93.
9. *Petroleum Economist*, April 1984, p. 146.

10. *Udenriksministeriets tidsskrift*, Copenhagen, February 1984, p. 47. In 1980 the exchange rate was $1 = FIM 425. In February 1985 one dollar equalled FIM 7.12; in September 1986 the rate is 5.18.
11. *Oil and Gas Journal*, 26 April 1982, p. 93.
12. This equipment was originally ordered from American companies, but these deliveries were stopped by the United States embargo on oil and gas technology imposed after the Soviet invasion of Afghanistan.
13. Egil Bergsager, 'Offshore the USSR', *Noroil*, January 1982.
14. Ibid.
15. TASS, 30 July 1982.
16. Ibid.
17. *Petroleum Economist*, April 1984, p. 146.
18. *Petroleum Economist*, April 1984, p. 141 and *European Offshore Petroleum*, 21 November 1984.
19. *European Offshore Petroleum*, 21 November 1984.
20. Sattarov op. cit.
21. *Petroleum Economist*, July 1985, p. 254.
22. *Oil and Gas Journal*, 12 November 1984, p. 64.
23. *Petroleum Economist*, April 1984, p. 146.
24. *Financial Times*, 25 July 1985.
25. *Oil and Gas Journal*, 4 February 1985.
26. Bergsager, op. cit.; *Petroleum Economist*, July 1985, p. 254.
27. *Hufvudstadsbladet*, 23 March 1985; *Oil and Gas Journal*, 15April 1985, p. 40.
28. Bergsager, op. cit.
29. Exhibition brochure, *Neftegaz-84*, Moscow, The Export Council of Norway.
30. *Petroleum Economist*, April 1985, p. 146.
31. *Arctic News Record*, August 1985, p. 32.
32. Sattarov, op. cit.
33. Source: Local officials in Murmansk, December 1984.
34. Matthew J. Sagers, 'National gas liquids and the Soviet gas processing industry', *CIR staff paper*, Bureau of the Census, Washington D.C., 1986.
35. *Sotsialisticheskaya Industriya*, 4 February 1986.
36. Cf. Thane Gustafson's account in *The Soviet Gas Campaign*, Rand, Santa Monica, 1983.
37. V. N. Ptitsyn came to Murmansk from the Leningrad party organization in 1956. He became chairman of the Murmansk regional party committee in 1971, candidate member to the Central Committee in 1976 and full member in 1981 and was re-elected in 1986. (Who's Who in the Soviet Union, 1984). The chairman of the executive committee of Murmansk town, V. I. Goryachkin,

who is known to be an able administrator, has publicly promoted Murmansk offshore interests even on Norwegian television.

38. The offshore research and production institute 'VNIPI morneftegaz' seems to play a central role in contacts with Western companies (see appendix 1).
39. Source: Tore S. Jørgensen: 'Miljødata av betydning for drift av anlegg i nordlige farvann' ('Environmental data of importance for operating installations in Northern waters') Nord-norsk petroleumskonferanse, Harstad, 27-30 August 1985.
40. Surveys carried out by the Norwegian Petroleum Directorate, NPD since 1973 (Egil Bergsager: 'Future Petroleum Potential of the Barents Sea', paper given at Habitat of Hydro- carbons, Stavanger 1–3 October 1985, p. 6).
41. In addition, Arktikugol, a Soviet mining enterprise, is drilling a second petroleum exploration well on Spitsbergen. An explicit aim is to understand more of the geology of the surrounding shelf. (Interview with leader of the drilling operation, on Norwegian television, September 1986.)
42. Finnish companies have indicated an interest in participating in a service base in Kirkenes. The base currently under development—Barents Base Kirkenes—is, however, owned by local interests.
43. *Report to the [Norwegian] Parliament*, no. 79, 1984–5, p. 5.
44. This conclusion corresponds very well to Robert Campbell's (op. cit.): 'past experience would suggest little possibility that the Russians can cope effectively with the technological demands of a big offshore effort of their own' (p. 157).
45. During his visit to Norway in July 1985, the Soviet gas minister V. S. Chernomyrdin confirmed that the number of drilling rigs would increase in the near future (*Aftenposten*, 2 July 85). The two giant jack-ups currently under construction at the Vyborg yard in cooperation with Rauma Reppola will probably be assigned to the Barents Sea after completion in 1986 or 1987.
46. In the Norwegian sector, they would probably have been called 'strategic blocks'.
47. Further details in *Gazovaya Promyshlennost* no. 6. 1984, p. 41.
48. *Gazovaya Promyshlennost*, no. 9, 1985, p. 5.
49. *Gazovaya Promyshlennost*, no. 12, 1985, p. 3.
50. *Gazovaya Promyshlennost*, no. 1, 1986, p. 8. According to the plan, the gas ministry is committed' to increase the efficiency per 1 meter geological-exploration drilling by 1990 1.4–1.5 times compared with 1985; increase the commercial drilling speed 1.4–1.5 times, total drilling per floating drilling unit not less than *2 times*, work productivity not less than 10 per cent. Reduce by half the growth rate of costs per 1 rouble produced.' (Idem.)
51. According to regional managers, December 1984, the number will increase to twelve by 1990.

3 Petroleum and military strategy in the Barents Sea: A conflict of interests?

In this chapter we examine the connection between Soviet military and petroleum interests in the Barents Sea. Our aim is to discuss the military-strategic framework for the search for hydrocarbons in the region; in other words, to what extent petroleum exploration and production may compete or collide with the operational needs of the Northern Fleet of the Soviet Union.

We must emphasize three points. First, the military interests of the Soviet Union comprise *the whole* Barents Sea, while its petroleum interests are mainly restricted to the *eastern part*, i.e. the continental shelf east of the 'disputed area', or a future borderline between Norway and the Soviet Union[1]. The shelf is part of the national territory as far as exploration and exploitation of natural resources is concerned, whereas the sea outside territorial waters is open to all. Activities on the Norwegian shelf might, however, be an impediment to the navigational scope of the Northern Fleet. From a military strategic viewpoint, the economic activities on the Norwegian and Soviet shelves are mutually dependent; in the economic sense the dependence exists only in such cases where an exploitable area extends to the shelf of the other side, or in cases where the parties agree to cooperate on exploration and exploitation.

Second, we must note that the various vessels of the Northern Fleet have different needs as to operational conditions in the Barents Sea. Surface vessels must mainly restrict their operations to the ice-free areas, which are also the main areas of economic activity. Submarines, however, can navigate underneath the ice, and can thus use a larger area. In other words, submarines can more easily avoid collisions of interest with civilian activities than can the surface vessels, as they are to some extent able to use alternative areas. On the other hand, it is commonly believed that during the last half of the 1970s, the Barents Sea was made a station area for the Northern Fleet's strategic submarines (SSBNs); this implies that submarines now remain in the area, thereby competing for space that civilian users in the long term might wish for their own activities. Consequently, in a longer time perspective, there might be an increasing risk of confrontation between civilian and military interests and requirements also as regards submarines.

Last, but not least, the two interested parties we here bring into focus are not the only ones. Shipping, the fishing industry, and researchers all have interests and demand their place in the Barents Sea. These interests also contribute to the 'traffic problem' in the region, and will be discussed when necessary to shed light on relations between military and petroleum interests.

Two kinds of factors restrict the scope of activities in the region: limitations caused by nature, such as ice and climate conditions, and the restrictions caused by the activities of other interested parties. Both types of factors will be considered here. Ice is the most important natural constraint for all groups involved.

Ice conditions in the Barents Sea[2]

In January, as much as 56 per cent of the Barents Sea is covered with ice, i.e. approx. 670 000 square kilometres. The ice-free area is then about 524 000 square kilometres. This proportion between ice-covered and ice-free sea remains more or less constant throughout the winter, from January till May. In September, which is the best of the 'summer' months, only 1–4 per cent of the sea is covered with ice: between 12 000 and 47 000 square kilometres. Surveys from 1971–80 show that in September one can sail all the way up to Franz Josef's Land[3] (see figures 3.1 and 3.2).

The ice-covered area consists of ice floes, ice ridges and icebergs of varying thickness and size, particularly different in expanse. Most of the ice is formed in the Barents Sea itself. There is, however, from time to time, a large supply of multiyear ice which comes from the Arctic Ocean, through the passages east of Svalbard;[4] the drift of ice goes in a southwesternly direction mainly caused by winds[5].

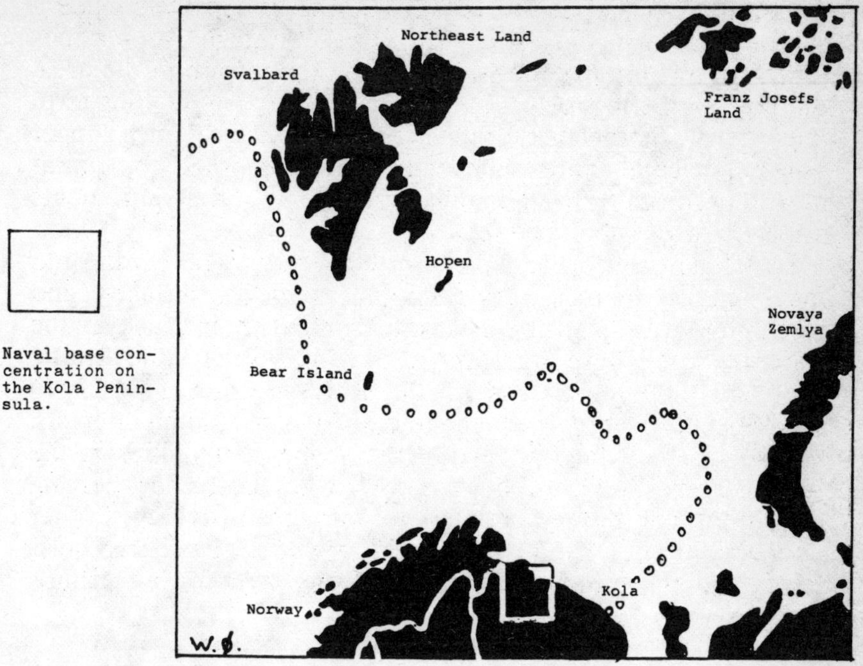

Figure 3.1 Maximum extension of ice in January based on measurements 1971, 1978–80

Observations made at Hopen indicate that 30–50 per cent of the ice-covered area is ice ridges. In extreme cases the ice ridges, which on average have ice keels of about twelve metres, constitute 70 per cent of the ice in the area; ridges have been observed with ice keels as deep as forty-eight metres[6]. In the winter density is one to six ice ridges per kilometre, while in the summer the average is nine per kilometre.[7]

Icebergs in the Barents Sea are relatively small; still, the largest can make furrows in the sea bed at a depth of one hundred metres[8]. Two areas have a particularly large number of icebergs: between Bear Island and Hopen, and between 30 and 60 degrees east (see figure 3.3). The icebergs calve from glaciers on Northeastland and Franz Josef's Land; at Hopen it is estimated that on average forty to fifty icebergs drift by in April, which is one of the worst ice months in the Arctic[9].

The Barents Sea is shallow, with an average depth of 230 metres. Of the area, 25 per cent (approximately 294 000 square

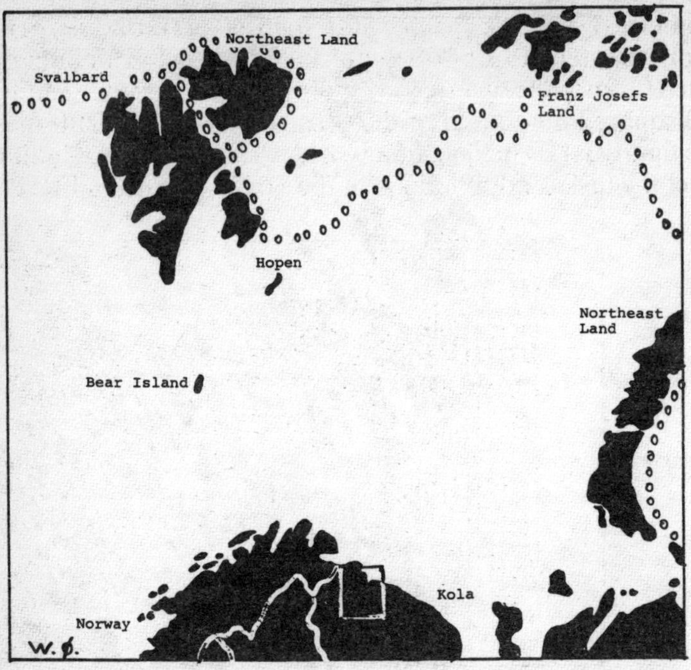

Figure 3.2 Maximum extension of ice in September based on measurements 1971, 1978–80

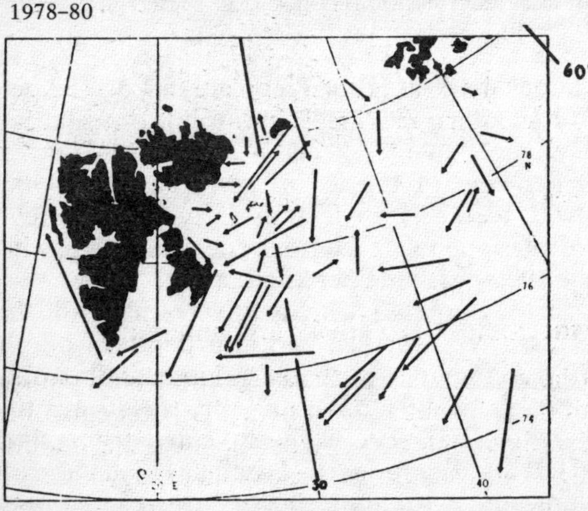

Figure 3.3 Ice drift direction and speed in the Barents Sea as determined from drift of ice floes and automatic buoys 1970–80

Source: Torgny Vinje, *The Western Barents Sea Physical Environment*, Norwegian Polar Institute, Publications, no. 179

kilometres) has depths of 300 metres or more; 9 per cent (approximately 112 600 square kilometres) has depths of less than 100 metres, whereas 67 per cent (i.e. approximately 800 200 square kilometres) of the area is deeper than 200 metres[10] (see figure 3.4). The question is: how do these natural limitations affect the tasks and functions of the Soviet Northern Fleet?

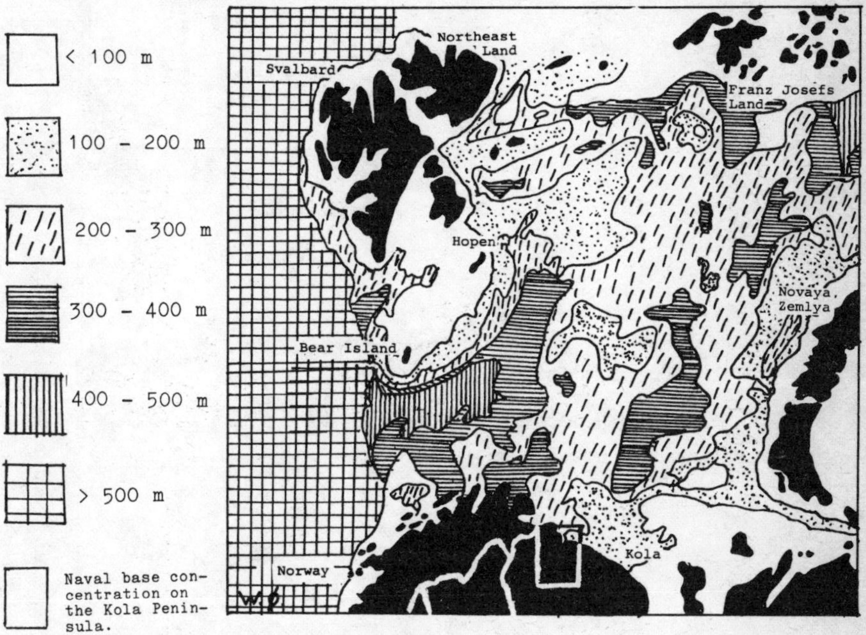

Figure 3.4 Bottom topography of the Barents Sea

The Northern Fleet: size, tasks and functions

The Northern Fleet has its bases on the Kola Peninsula in the area from the Pechenga Fiord near the Norwegian border and eastwards. Severomorsk, which lies at the mouth of the Murmansk Fiord, is the main base of the fleet, whereas Polyarny, further into the same fiord, is the main base for strategic submarines. All in all, there are twelve naval bases on Kola, ten of which lie within a very limited area, close to the seaward approach to Murmansk (see figure 3.5). The main bases are thus

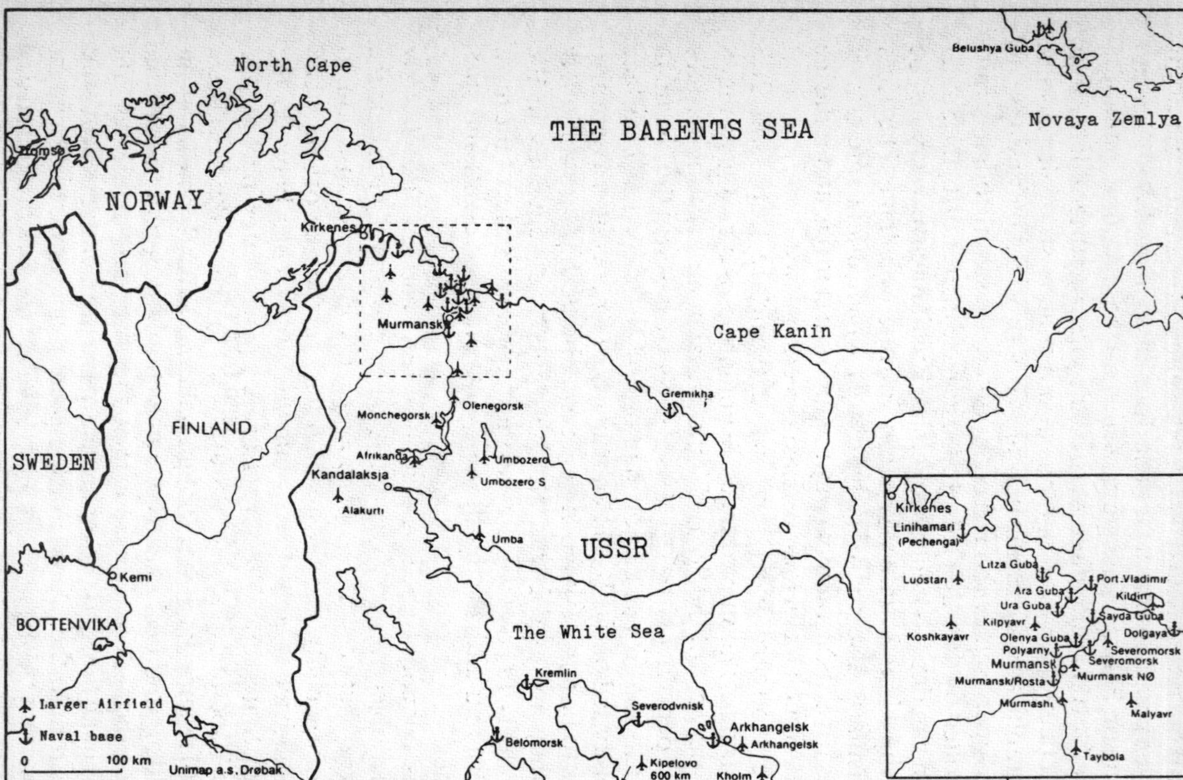

Figure 3.5 The naval base and military airfield concentration on the Kola Peninsula

Source: *The Military Balance, 1985–86*, Norwegian edition, the Norwegian Atlantic Committee, p. 140. Reprinted with permission

centred on a coastline which has access to an ice-free sea the whole year round. Only one base on the Kola Peninsula, Gremikha, is located so far east that it has ice during part of the year[11], a problem which faces the four bases in the White Sea and the one on Novaya Zemlya as well.

The Northern Fleet seems to be the most important of the four Soviet fleets for both geographical and military strategic reasons. Geographically, this is the only area where Soviet battleships can reach open sea without being forced to sail through straits controlled by the other side. From its base in the north, the Soviet Union can deter potential adversaries with nuclear weapons, be they the United States, Western Europe or China. The Kola Peninsula is also the only base area which offers safe transit routes for SSBNs on their way to and from stations under the ice in the Arctic Ocean and the East Greenland Sea.

The Northern Fleet now has 220 surface ships and 203 submarines. The importance of submarines is striking: as much as 57 per cent of the Soviet Union's total number of submarines and 52 per cent (41 of 79) of its SSBNs are based in the North. In contrast, the Soviet Union has only 220 of its 1277 surface vessels based on the Kola Peninsula, i.e. 17 per cent[12]. There also seems to be a pattern when new Soviet battleships are stationed: first they are used in the Northern Fleet, subsequently in the other fleets. In this way, the Northern Fleet has some of the most modern and effective battleships, both surface vessels and submarines.

It is commonly believed in the West that this fleet has six distinct objectives: (1) To contribute to maintaining the balance of terror between the superpowers; (2) in a war situation to destroy or neutralize Western aircraft carriers before they can attack Soviet targets; (3) to cut off supply lines between the United States and Western Europe; (4) to offer backup for landing operations and for the army; (5) to defend the Kola base, and (6) to support Soviet foreign policy. A seventh objective can be added: to protect transit routes within the Barents Sea (from here on termed internal transit), so that the SSBN force has safe access to any extended Arctic station area, including beneath the ice in the Arctic Ocean[13]. This point, which must be seen in conjunction with goal number 1, has become particularly important in view of the American forward-based naval strategy, which involves fighting the Soviet Navy in its own waters in the opening phase of a war. Sealing off the Barents Sea against enemy intruders must therefore be a high-priority goal.

In order to achieve goals 2, 3, 4 and 6, it is vital for the Northern Fleet to be able to move out of the Barents Sea to the war zones (henceforth termed 'external transit'). The Barents Sea thus meets an important need for the Northern Fleet as regards external transit. The Barents Sea also seems to contribute to the balance of terror in that it apparently has become a strategic station area. This involves patrols through and underneath the ice, both in the Barents Sea and in the Arctic Ocean. To achieve goals 1, 5 and 7, regional stationing of the appropriate naval units is required. This again means that the Barents Sea also must be used as an area for military exercises of the Northern Fleet. The question is whether these four kinds of use: external transit, internal transit, stationing and military exercises are feasible, given the limitations set by nature and other restrictions. Let us first examine the situation of the surface fleet.

Operational needs and conditions of the surface fleet

The surface fleet must operate both within and outside the Barents Sea in order to achieve its part of the seven goals assigned to the Northern Fleet. If we disregard the point concerning military exercises, surface ships will primarily have tasks involving the solution of points 3, 4 and 6, i.e. transit. One part of the surface fleet will in addition have assignments involving protection of SSBNs which might be in operation both within and outside the Barents Sea, i.e. anti-submarine warfare missions (ASW).

The ice-free minimium area which can be used by the surface vessels on a whole-year basis is, as previously mentioned, somewhat more than 500 000 square kilometres. Parts of the marginal ice area may also be used for military tasks (strategic ASW) by nuclear powered ice-breakers stationed in the region. Since the Soviet military must be capable of solving its assigned tasks at any time of the year, this is the area for which all war operations must be planned. For the Northern Fleet, this is the worst-case situation. Within this area the fleet will have to master both transit and regional operational tasks in a war situation. Let us look at the problem of transit first.

In January, with maximum ice formation, the average distance from Kola to the ice edge is approximately 330 kilometres. The passage between Bear Island and northern Norway is about 430 kilometres wide, which means that the ice-free operational area expands westwards. More space would naturally be preferable to permit greater dispersion and less channelling of the vessels to areas vulnerable to enemy counter attacks. On the other hand, there is little to be done with this nature-determined channelling effect, and the Northern Fleet has lived with it for a long time.

From the viewpoint of the Soviet military, any further limitation of this operational area is not desirable. There are, however, other activities which demand space thereby reducing the operational scope. For instance, extensive fishing takes place in the Barents Sea, in terms of quantity of fish catches, and number of fishing boats. Year-round fishing for cod and haddock goes on in that part of the Barents Sea which is usable as fairway for the Northern Fleet. Trawl and longline are the fishing methods used, and these both demand more room than the space occupied by vessels singly or in groups. In peacetime, a certain competition for available space can occur. This would hardly be the case in wartime when the fishing will simply have to adjust to the military strategic requirements. This also applies to shipping and research activities. Oil rigs, however, cannot so easily be removed in a war situation. They can be blown up, but this means risking serious blow-outs, which in turn might interfere with war operations. If left behind, oil rigs will occupy space in competition with battleships.

Soviet authorities are faced with different situations on the Norwegian side and the Soviet side. On the Soviet shelf they can plan use of the available area so that civilian and military activities can exist side by side. It also appears that geology is to some extent favourable for Soviet military interests: a large part of the exploration has, as described above, taken place in the eastern and southeastern part of the Barents Sea (see figure 2.1). However, petroleum activities at Murmansk High and Severo-Kildinskaya may create obstacles to the passages used by surface vessels.

Soviet authorities cannot plan exploitation of the area on the Norwegian side of the Barents Sea; here they have to adjust to Norwegian plans and decisions. In the years to come the entire area between the Norwegian mainland and Bear Island is to be opened up for oil exploration, with foreign participation. This implies that the whole passage area for the Northern Fleet on the Norwegian side will be affected by activities which in the long run may lead to oil development.

Economic terms, not strategic considerations will be decisive for Norwegian authorities. A conflict may thus arise between the military needs of the Northern Fleet to reduce the channelling effect and Norwegian petroleum interests. The geological conditions determine where the rigs are to be placed—not concern for Soviet security interests, even if Norway has paid attention to these interests in the post-war period in many other respects. The decisive question in this context concerns the number and location of installations, and the distance between them. A 'forest' of rigs in the most important fairways will cause problems for the Northern Fleet, and might cause incidents and conflicts between the USSR and Norway. It is clear, however, that oil development in this area, no matter what the density or location of rigs, will increase the channelling effect for surface vessels. From a Soviet military viewpoint, any kind of narrowing down of its operational area, already severely limited by ice, is not desirable. That does not mean that the Northern Fleet cannot live with some narrowing down of the fairway. What counts is the degree of channelling, and where it takes place. It is worth noting here that the ice-free area on the Norwegian side is larger and wider than on the Soviet side, and that space for development is therefore more ample. Furthermore, the channelling effect can be reduced if the oil fields on the Norwegian side are developed partially with production equipment on the seabed.

The Soviets may feel psychological discomfort at the prospect of foreign rigs permanently placed close to Soviet military areas. They may also see a possibility of exploiting the presence of the rigs for political purposes, under certain circumstances and situations. There are several examples of Soviet overreaction to Western presence and activity in Northern areas, especially on Svalbard[14]. Suspicion of Western intentions seems to be a rather common feature in Soviet behaviour in the area[15]. Such an attitude may have been expressed through breaches of treaties which the USSR itself has agreed upon. In direct defiance of the Norwegian–Soviet Grey Zone Agreement of 1978 Soviet vessels have boarded foreign fishing boats with a Norwegian licence, fishing in the area. According to the agreement, this is a

Norwegian responsibility. It is also worth noting that, as far as fishing is concerned, the presence of fishing boats is only temporary, whereas in the case of oil production the presence will be permanent. With this in mind, it is probable that the Soviet reaction will be especially negative to foreign-owned rigs on the Norwegian Shelf. Furthermore, the Soviet Union may find it in its interest to exploit the situation politically, for instance by demanding concessions from Norway in other areas.

The potential use of oil installations for military purposes is rather limited. Because of the noise they produce, they are unsuitable as listening posts using passive sonars. Due to the danger of explosion, they cannot be used for launching missiles. In a war situation, the platforms will probably be evacuated as early as possible, out of consideration for crew safety. During the Second World War, the population on Spitsbergen was evacuated since their security and food supplies could not be guaranteed. It is even harder to offer such guarantees to those onboard platforms which might even be located close to probable war zones. The visual detection of fleet movements etc. which could be made from a rig has little military value compared to the information that can be gathered with other intelligence methods. The main problem for the surface fleet is thus not so much what military use the West can make of oil installations, but the fact that the rigs channel Soviet external transit to easily identifiable areas.

The pattern followed by the Northern Fleet in its military exercises reflect the operational tasks it is meant to solve in wartime. The concentration of strategic submarines in the Barents Sea implies that part of the surface fleet has been assigned ASW-tasks for times of peace and war; it also indicates that this fleet has main responsibility for the defence of the base complex on Kola.

The great emphasis on anti-submarine warfare can be deduced from the composition of the fleet: about 70 per cent of the USSR's ASW Kresta-II cruisers and 75 per cent of the ASW attacksubmarines, the ALFA, TANGO and VICTOR-class, (equipped anti submarine missiles and torpedoes) are all based on Kola. This deduction is also valid in the case for the ASW-aircraft carrier Kiev and the combatants of Kirov, Udaloy and Krivak classes. The ALFA submarines are the fastest and can dive the deepest of any attack submarine ever built. There is a great ASW-preparedness in the Northern Fleet, and the effort put into ASW-tests in the 1970s and 1980s indicate the importance of this function.[16]

ASW-forces usually have imposed operational restrictions in or near a station area. Such forces will normally not be given permission to attack submarine contacts in a station area to avoid mistakes and hence attacks on one's own submarines. In other words: the ASW-forces operating in the Barents Sea must impose operational restrictions on themselves as long as this functions as a station area. In an area where the highest priority is the defence of the country, i.e. the protection of the base complex on Kola, it must be hard to accept restrictions on offensive actions. Hence, there seems to be a certain contradiction between some of the presumed tasks of the Northern Fleet. None of these tasks can be handled in the best possible way as long as they all have to be executed in the same area. The question is then whether the Northern Fleet is forced to make such compromises, or if one of these tasks can be attended to outside the Barents Sea without any disadvantages. We return to this question later.

Submarines and petroleum activity

Of the seven main assignments of the Northern Fleet, submarines will mainly contribute to maintaining the balance of terror, of attacking aircraft carriers and the supply lines between the United States and Western Europe. The first task can theoretically be ensured by operations *inter alia* in the Barents Sea; the two other tasks require transit out of the region to areas south of the Greenland-Iceland-United Kingdom (GIUK) gap. Stationing and transit, both external and internal, may therefore be the submarine fleet's primary use of the Barents Sea, including military exercises.

Ice has a different significance for submarines than for the surface ships, since the former can expand their operational space by going under the ice. Here the submarine crew is faced with other limitations: the keels of the icebergs and pack ice can be so deep, up to 100 metres, that there is no room for the submarine between the ice and the bottom. The question here is which depths are needed to meet safety requirements and to complete the task? The answer will differ for SSBNs and for attack submarines (SSNs). Let us therefore discuss separately the operational conditions for these two main types.

The Barents Sea as stationing area for SSBNs

It is now usual in the West to see the Barents Sea as part of a larger strategic station area in the Arctic. This has not always been the case: in 1981 the Pentagon identified the Barents Sea as the one and only Arctic station area[17]; three years later the same source expanded the area to include the Barents Sea, the East Greenland Sea and the areas north of the GIUK gap[18]. Simultaneously with this geographic expansion came another: American Navy authorities in Washington acknowledged that the Soviet SSBNs show great interest in operating under the ice of the Arctic Ocean, in the marginal ice zone. This includes ice with a thickness of up to three metres, which the large SSBNs can break through using their own buoyancy[19]. The average thickness of ice in the central Arctic Ocean is three metres, the same as in the marginal ice zone. Before looking at the background for this geographical expansion of the station area, it might be interesting to study the qualities of the Barents Sea as a station area in itself.

Nature-determined conditions for SSBN-operations in the Barents Sea

The Barents Sea offers both advantages and disadvantages as a station area. The advantages are obvious. First, the distance is short between home base and the operational area. In the 1960s, Soviet SSBNs spent as much as two-thirds of their operational time at sea in transit between base and station, between Kola and the station area outside the American east coast; today transit time has been reduced to an absolute minimum for the modern SSBNs, Delta and Typhoon. In the north, base and station areas are approximately identical. Secondly, the station lies in an area in which the Soviet Union has nearly complete military control, both at sea and in the air. This provides the best conceivable protection for the SSBNs. Sonar range is also drastically reduced since the station area lies in shallow coastal waters. As sonar range can be reduced to only a few nautical miles in such waters, this means that advancing Western attack submarines will have great problems tracing Soviet SSBNs.

The ice can also be used as a hiding place for these vessels; in the transit area between the ice and open waters there is a lot of noise which the SSBNs can hide in. Passive sonars and the equipment used to separate submarine sounds from the background noise of the sea will not be very effective in this area. If the SSBNs choose to go further in under the ice, they will also bar the enemy's possibility of detecting them with ASW-equipment other than attack submarines[20]. In this way they can use the areas under the ice as a 'protected sanctuary' for strategic patrols.

It must be added that Soviet submarines operate under the same sonar conditions as the attacker. The Soviet forces will therefore have great problems discovering and tracking down submerged attackers trying to sneak into the area[21]. In this way, Soviet counterattacks are rendered more difficult. It is a fact, however, that it is easier to hide an SSBN than to find it; SSBNs are defensive in their self protective role, and will have to retreat to, and operate in, waters which can give them protection. They can also lie quietly on the bottom and reduce their own engine and reactor noise. In this position they are almost impossible to find. An attacking SSN will not be as familiar with local operational conditions as the Soviet vessels are. Western SSNs therefore have an operational handicap compared to the vessels they chase: the attacker must in addition protect himself from possible attacks from Soviet ASW-forces. All in all, the SSBN fleet has more to gain from local conditions than has the attacker[22].

The noise conditions in the Barents Sea are also of great importance in this context. Fishing activity, exploration for oil and gas, the local shipping trade, seismic research on land and in the sea, animal life in the sea—all create background noise, complicating detection by passive sonar. The computers aboard the vessels and on land are not able to separate submarine noise from the general background noise of the sea which provides cover for a submarine. Compared to the North Sea, the Barents Sea is certainly a relatively quiet area; but there is still enough noise to complicate detection. From this viewpoint, it is the SSBNs which have the relative advantage of the background noise. This is interesting when considering the exploration for petroleum on the Soviet Shelf. If this results in production in the future, the Soviet Union will be able, in a possible war situation, to manipulate the noise level in the sea by activating and deactivating the production equipment. Even if oil production itself should stop, the equipment can still be used for producing noise. In this way the oil installations can, for a limited period, be used for military purposes. In those periods when the equipment does not produce noise, passive and active sonars installed on the rigs can be used for detecting Western submarines.

This will primarily apply to the production areas in the westernmost part of the Soviet Shelf. The search areas in the southeastern corner of the Barents Sea are so shallow, and lie so far away from the attack routes of Western submarines, that they will hardly have any significance for detection purposes. A similar alternative use of platforms on the Norwegian side is also possible. In this case, however, it is viewed as less necessary because Soviet submarines are generally more easily detectable due to self-generated noise than are their Western counterparts. The Soviet Northern Fleet thus has a greater need for noise manipulation and additional sonars installed on platforms than has the West, but development of increasingly quiet-running Soviet submarines can in the long run alter this situation.

With this in mind, we suggest that these supplementary military functions of oil platforms will probably have a relatively marginal military value both in the East and West. It seems fair to conclude that SSBNs operating in shallow waters, close to the coast, with a lot of noisy activities going on, have the best possibilities of not being detected. This makes the Barents Sea a favourable and suitable station area, but not without disadvantages: the USA will have longer warning time for missiles launched from the Barents Sea than from the area off the American east coast. The combination of ice protruding deeply into the water, and shallow waters will in some areas prevent SSBN traffic in the winter. This poses problems for the submarine commander; he must first decide which depths he needs in the station area to have sufficient space to manoeuvre in a submerged position; second he must calculate how much room the submarine needs between the ice and the bottom in transit from one area to another. Safety of the vessel and the crew has to be considered, and also the reliability of the weapon system, which can be reduced in contact with ice.

It is not possible to give an exact, universally valid depth in which a SSBN can manoeuvre in a station area; the depth depends on a series of local conditions to be evaluated in each individual case. On the other hand, this does not mean that there are no rules of thumb. Based on their general experience, Norwegian submarine captains have found this depth to be around 200 metres[23]. Independent Canadian estimates generally confirm this suggestion, calculating the necessary depth to be 183 metres[24]. For this reason, the Canadian security analyst, Professor Harriet Critchley, estimates that about half of the Barents Sea does not offer sufficient depths for safe SSBN operations[25]. If we take 200 metres as the depth necessary to manoeuvre SSBNs sufficiently for them to fulfil their functions in the station area, this means that about 33 per cent of the Barents Sea (about 395 000 square kilometres) cannot be used for such purposes on a year-round basis. The space between the ice and the bottom will be too narrow. These areas can be found in the northwestern corner of the Barents Sea, from Northeastland in the north and southwards past Bear Island; depths here are less than 100 metres for a continuous stretch. Since this is where the greatest concentrations and most frequent occurrences of icebergs and pack ice are located, this area is unsuitable as a year-round station area; this is also the part of the Barents Sea in which one finds furrows in the sea bottom caused by icebergs as much as 100 metres deep. Nor does the area meet the requirements for operational depth in summer.

The other large area of the bottom where depths are less than 200 metres can be found in the east, along the coast of Novaya Zemlya. As is indicated in figure 3.6, this leaves an area in the central Barents Sea of about 800 000 square kilometres which has depths between 200 and 500 metres. Not even all this area meets the demands for full manoeuvrability; about 63 per cent of the area (506 000 square kilometres) has depths between 200 and 300 metres. This implies that SSBNs have to face unfavourable operational conditions in large parts of these areas; between 30 and 60 degrees eastern longitude deep icebergs are not uncommon, which complicates safety in the winter. In this context, the size of the vessels mentioned is important; Delta submarines are between 140 and 150 metres long and weigh from 9000 to 10 500 tons when submerged[26]. (Much longer than a football field (110 metres) the Typhoon is in fact as tall as a ten to twelve storey building and wider than a handball field.) It should be unnecessary to point out that such vessels need considerable space to manoeuvre, both in the horizontal and vertical plane. From the end of the 1950s, there have been reports of submarines which have sailed into hollows in the Shelf without any outlet to the deep sea, and have had difficulty getting out

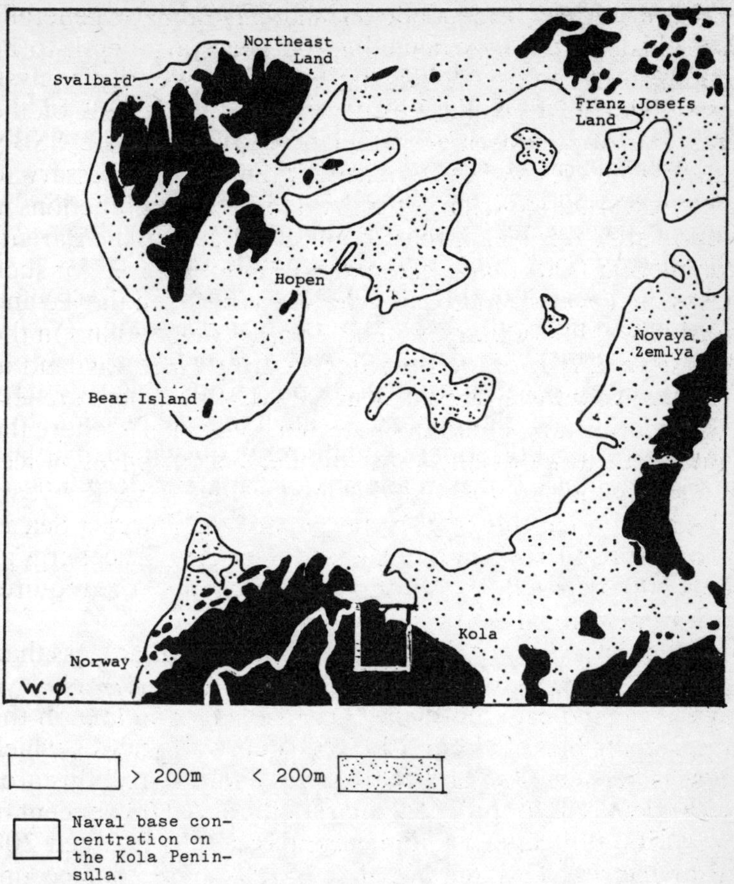

Figure 3.6 The 200-meter depth contour

because of ice conditions above the ship - the so-called 'valley of death' problem. Similar seabed conditions can also be found in the area between Northeastland and Franz Josef Land, where there is also a high frequency of icebergs. Clearly, ice conditions can in such cases be a question of life or death. One does not take unneccessary risks in a station area. It is only with a depth of 300 metres or more that there is the necessary operational space between the ice and the bottom, and horizontally between deeply protruding icebergs.

On the other hand, icebergs are not so frequent, as to make it impossible from time to time to operate in certain shallow waters, even in winter. There is, however, reason to believe that the Soviet Union is careful in this respect and limits such missions, in order not to endanger ship and crew unnecessarily. In this way these waters can be considered as the stationing area's marginal operation zone: one can operate freely there in the summer, but must impose limitations in the winter (see figure 3.7). Only 25 per cent of the Barents Sea (i.e. about 294 000 square kilometres) have depths of over 300 metres or more. It is probably necessary to have such depths to be able to operate under the ice in the station area. All in all this is equivalent to an area of about 163 000 square kilometres in the winter. This area is not contiguous, but is made up of separate deep hollows

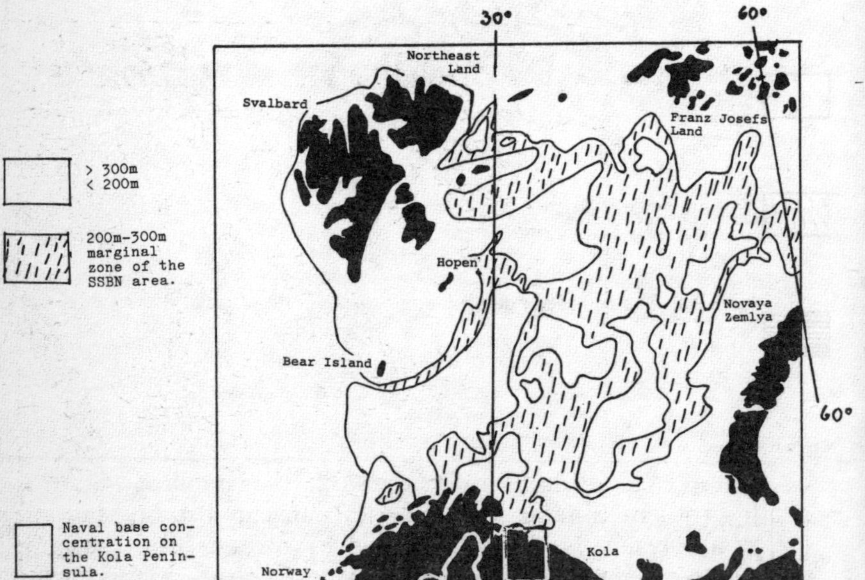

Figure 3.7 The marginal zone of the SSBN-stationing area in the Barents Sea

connected with depths of 200 metres or more. This continuous stretch of 200 metres' depth is what we have termed the stationing area's marginal operation zone. The Northern Fleet thus cannot count on being able to use these parts throughout the year, on a complete and unlimited basis. The remaining 131 000 square kilometres with sufficient depths under the ice, are to be found in the southern and central part of the Barents Sea, with direct access from the ice-free area, which is estimated at around 150 000 square kilometres. Depths are here 200 to 300 metres. The total extent of an all-year stationing area where manoeuvres can be performed safely will be around 281 000 square kilometres (see figure 3.8). The question is how many SSBNs can operate in this expanse of sea ?

The problem of the size of each vessel's manoeuvre-space is complex. It depends on many factors, like the natural characteristics of the region, the submarine's qualities, the range of

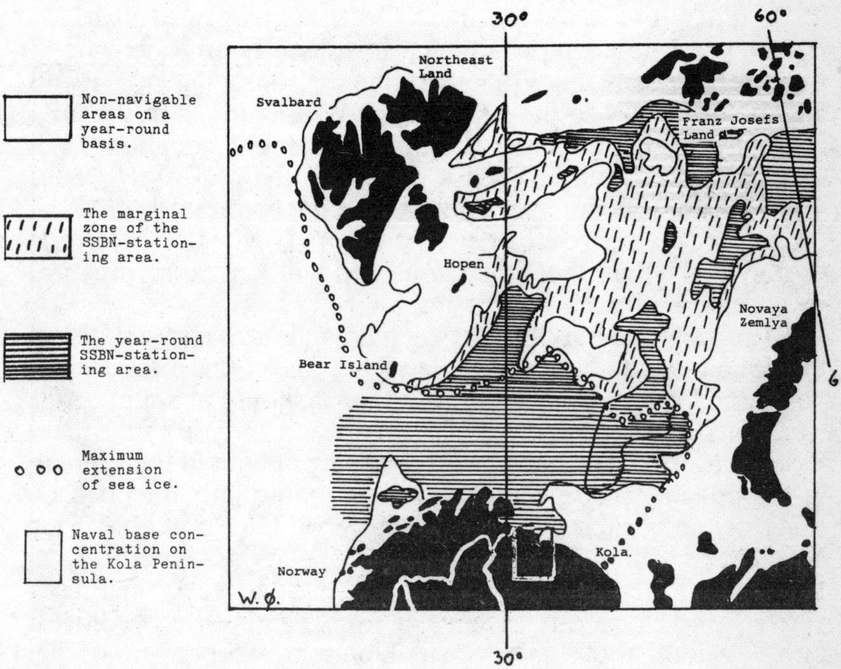

Figure 3.8 The year-round stationing area in the Barents Sea

weapons, sensor capacity, etc. A highly competent submarine authority in Norway estimates that each vessel as a minimum will need an operational radius in the horizontal plane of 50 times 100 nautical miles—100 times 100 nautical miles, or approximately 17 000—34 000 square kilometres[27]. This will provide sufficient space for a total of eight to sixteen SSBNs. Today the Soviet Union has twenty-six modern SSBNs stationed in the North—twenty-two Deltas and four Typhoons. Theoretically, 30–60 per cent of the SSBN-fleet can find a station in the Barents Sea[28].

This kind of conclusion is only of theoretical interest. In reality the Soviet Union will have neither the capability nor the wish to place so many submarines in such a limited sea area, for many reasons. First, most of the Soviet submarine fleet at any given time is in port for repairs, maintenance and bunker; American estimates indicate that only 15 per cent of the Soviet SSBNs are out of base simultaneously, which means that at the maximum three to four vessels are in the stationing area at any given time. Moreover, the SSBNs are not alone in the area; there are also SSNs, ninety-five in the Northern Fleet, and they need space for transit, operational and training purposes[29]. Even if defence of the SSBNs mostly takes place outside the stationing area, it will also be necessary to have limited ASW-operations in the area itself. Professor Critchley has described the situation as a 'strategic traffic jam'[30]. Such a concentration of vessels would also give the enemy a possibility of concentrating his ASW-forces for a collective effort in a limited area of the Barents Sea. It is not likely that the Soviet military will find this situation favourable for them.

The part of the SSBN-fleet which lies in port is also of interest in this context. It must be assumed that some of these vessels can be readied relatively quickly if the need should arise; in a war situation one would wish to get them out and away from the main bases as swiftly as possible, to avoid being put out of the running in the opening phase of hostilities. Only an intact fleet can be used for the purposes ascribed to it by the US Commander for Naval Operations, Admiral James D. Watkins: '[the Soviets] keep their [SSBN] forces under the ice . . . to keep them as a viable force which after a nuclear exchange [would become] a key bargaining tool'[31]. The Soviet Union has 85 per cent of this fleet in port at any given time. If this part of the fleet is to play the role assigned to it by Admiral Watkins, it must come free of the harbours, which will be high-priority bomb targets in the early phase of a war. The Barents Sea is the closest and most natural recipient for the SSBNs of the Northern Fleet. If this stationing area is already filled with SSBNs in operation, there will be less operational space, and the result will be further traffic chaos. It is therefore reasonable to believe that the SSBNs in operation first of all will leave the Barents Sea and station themselves in the area north of the GIUK gap and under the ice in the Arctic Ocean. The Barents Sea is thus primarily to be considered a crisis and reserve stationing area for the port-based part of the SSBN fleet.

On the other hand, all vessels must in peacetime gather operational experience in those areas which can become their stationing area in a war situation. It is thus reasonable to assume that all the SSBNs in the Northern Fleet gain operational experience in the Barents Sea, but that at any given time there are fewer than the total number of those in operation. In this way the use of the Barents Sea in peacetime is toned down, whereas this function can have increased significance in wartime; all in all it therefore seems that operative SSBNs primarily use the Barents Sea for transit to other stationing areas.

If this assumption is correct, the Northern Fleet has reduced its problem of defending the Kola base. As previously mentioned, there will be less self-imposed limitations if the Barents Sea does not have a primary function as a stationing area. Dispersion of SSBNs to a larger Arctic area will reduce the problems with which the Northern Fleet otherwise would be confronted with because of all its different tasks.

SSBN-operations and petroleum exploration

Since submarines can sail under the ice, the operational area for SSBNs is greater than for the surface fleet. Besides, SSBNs require relatively great depths, whereas exploration for the time being is going on in rather shallow waters. Still, there have been incidents between Norwegian petroleum activity on one hand and personnel in the Northern Fleet on the other. On 11 July 1985, a Soviet frigate cut off the seismic cable of a Norwegian research vessel called *Marlene Østervold* fifty kilometres of the sector line in the Barents Sea. Less than a week later, Soviet authorities apologized for the incident. The explanation was that the frigate had unintentionally crossed the course of the *Marlene*

Østervold, as it had not picked up the radio signals from the boat. These apologies were positively received by the Norwegian government who interpreted them as confirmation that this was to be an isolated episode. The explanation as such was, however, harder to accept; the Norwegian crew had made several efforts to contact the two Soviet frigates which had been following the research ship, both by calling them up repeatedly over the radio, and by firing off a warning rocket, but in vain. It has therefore been asserted that the incident might have some connection with a Soviet naval exercise going on at the time; the noise from the seismic equipment may have disturbed underwater operations in such a way that the Soviet commander-in-chief chose to take action. If this explanation is correct, it illustrates the conflict of interests between military and civilian activity in the area.

It is also possible that the incident might be connected with an internal dispute between military and civilian authorities in Moscow, and that this was a military reaction to hints of co-operation with Norway on exploitation of petroleum resources in the area. With this in mind, Norwegian State Secretary Torbjørn Frøysnes said that the timing of the incident was very unfortunate, as it might have negative repercussions on such proposals of cooperation[32]. If an internal dispute between Soviet authorities were taking place, it is noteworthy that the apology was quite unreserved and expedient for the Soviet Union, and that the Soviets compensated for loss of the cable to the Norwegian company using the *Marlene Østervold*. The outcome indicates that the Soviet Union is still interested in petroleum exploration in the area, and that Soviet military authorities will have to be prepared for this. No matter how this incident is explained, it illustrates that problems of co-existence can arise between different interest groups, both internally in the Soviet Union and across frontiers. Not until one of the parties finds a large exploitable area will this conflict of interests be put to the test. For the present, it seems to be Soviet policy to accept both interest groups operating in the region.

There seems to be no insurmountable obstacles to co-existence between SSBN operations and petroleum exploration and production in the Barents Sea. The problem is more accentuated in connection with the surface fleet, which has greater operational limitations because of the ice. Indeed, our study indicates that there might be greater conflicts between some of the presumed tasks of the military, than between military and civilian interests. The conflict between the need for an offensive defence of Kola, and the necessary limitations to be placed on this defence in the stationing area, is in itself so important, that it is worth questioning the view of the Barents Sea as a vital stationing area for the Soviet fleet. The combination of many factors underlies this doubt:

- the limited operational area due to deeply protruding ice and shallow waters;
- the need to complicate the ASW-operations of the adversary;
- in a war situation, the need to secure space in the Barents Sea for the port-based part of the Soviet Northern fleet;
- the need for offensive defence of the Kola base with a point of departure in the Barents Sea.

It thus seems that the primary function of the Barents Sea for the SSBNs is still external transit. The Barents Sea appears to be only in moderate use as a stationing area in peacetime. In wartime this can change, if some of the port-based SSBNs seek to transit out of the area to the Arctic Ocean, or the Northern Atlantic if time permits. The question which remains to be answered is which operational demands have to be made for transit in ice-covered waters.

Transit conditions in the Barents Sea

SSBNs need less space between the bottom and ice when in transit than when stationed. This was clearly shown in the period 1957–62, when the American Pacific and Atlantic Fleet carried out protracted operations under the ice in very shallow waters. In January 1960—the year's worst ice-conditions—the USS *Sargo* sailed in a submerged position through the shallow and ice-covered Bering Strait and into the Arctic Ocean; for thirty-one days it sailed a distance of 6000 kilometres in the Bering Sea and the Chuckchi Sea, which has an average depth of 80 metres. In some places the boat was only seven metres above the bottom to avoid top collision with the ice[33]. Navigation with such small margins naturally assumes adapting the speed accordingly: average speed of the *Sargo* was only 4 knots. If the speed is increased, the room for manoeuvre must also be increased.

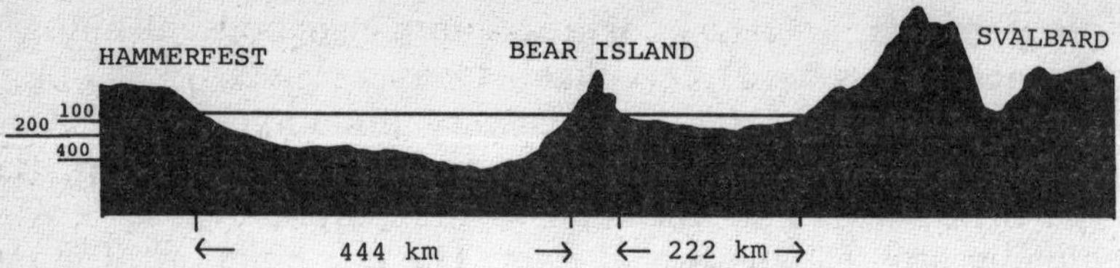

Figure 3.9 Cross section of depths between Svalbard, Bear Island and north Norway

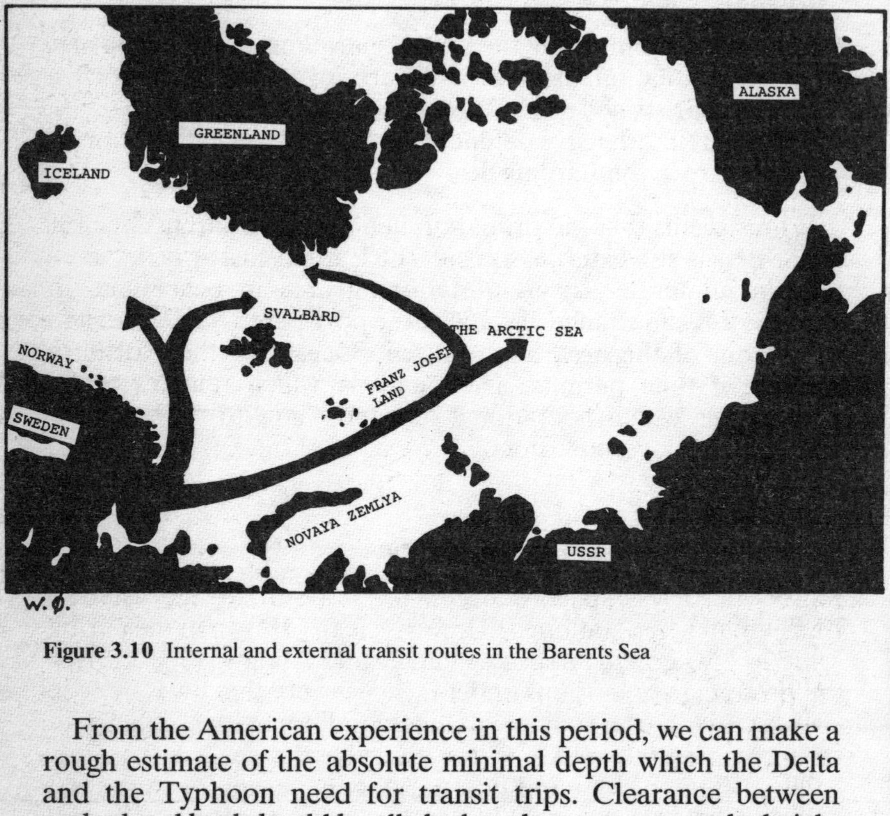

Figure 3.10 Internal and external transit routes in the Barents Sea

From the American experience in this period, we can make a rough estimate of the absolute minimal depth which the Delta and the Typhoon need for transit trips. Clearance between seabed and keel should hardly be less than ten metres; the height of a Typhoon from the keel to the highest point of the tower is estimated at about twenty-eight metres[34], whereas clearance between the submarine tower and the keel of surface vessels or ice ridges is estimated at about fifteen metres; the keels of boats or of ice ridges protrude about fifteen metres under the surface. All in all, this gives a minimum depth of sixty-eight metres. We can assume that the crew will scarcely find this minimum depth very comfortable, if they have to navigate with such limited margins for a long time, and especially if speed has to be increased. Preferred depth is hard to estimate, but we can safely add on about thirty metres, so that a relatively comfortable transit depth for the large SSBNs will be about one hundred metres.

In this sum, no allowance has been made for icebergs. This is because SSBNs, for reasons discussed previously, need to avoid areas with a high frequency of icebergs and shallow parts. This means that SSBN transits in the shallow areas between Spitsbergen, Hopen and Bear Island will be very risky, and are almost irresponsible in the period from January to May-June. The safest transit route on a year-round basis is between Bear Island and northern Norway. (see figure 3.10). From there on the vessels can choose between stationing in the Norwegian Sea, or continuing northwards through the Fram Strait and into the Arctic Ocean. Similarly, the sailing depths from Kola and northeastwards towards the outlet of the Arctic Ocean between Franz Josef's Land and Novaya Zemlya offer sufficient transit depths (see figure 3.4). Even if the frequency of icebergs here can be great, depths are sufficient to ensure a sailing room of one hundred metres. In the 1950s three approaches to the Arctic Ocean were described in the Soviet press: between Svalbard and Greenland, between Svalbard and Franz Josef Land, and between Franz Josef's Land and Novaya Zemlya (see figure 3.10)[35].

In the late 1950s and beginning of the 1960s SSNs were so small that it was possible to transit between Svalbard and Franz Josef's Land without too much risk; the smallest can probably still do so. It is, however, very doubtful if large SSBNs can copy such voyages on a regular basis today; this is primarily due to the seabed being quite hilly, so that the submarine must weave in and out of twisted valley formations in the subsurface, at the same time as there is a high frequency of icebergs. In this area submarines can get caught between the ice and the valleys—'the valley of death-problem'.

The safe transit routes are thus between Bear Island and northern Norway and between Franz Josef's Land and Novaya Zemlya. These transit routes are vital for the Northern Fleet to protect in order to maintain its strategic deterrent effect. The significance of this task has been accentuated as a result of the forward strategy of the American fleet, which in theory presupposes venturing into the Barents Sea. The Northern Fleet must prevent this eventuality with a major effort of its air force and navy in the areas bordering on the Barents Sea, to the north, south and west.

In this context too, the ice and depths contribute to channelling transit out of the Barents Sea, thus implying that the enemy can group and gather his ASW-forces and counter offensives around the outlets. Such a gathering of forces conforms with the train of thought behind the American fleet's forward strategy. This represents a military problem for the Northern Fleet, caused primarily by nature, not by people, but economic activity can worsen the channelling effect, as is the case for the surface vessels, and thereby improve the working conditions for the adversary's ASW-forces.

SSN-operations in the Barents Sea

The primary tasks of Soviet submarines are to attack the supply lines between the United States and Western Europe, to carry out attacks on aircraft carriers and to protect the SSBNs and Kola. The two first tasks require transit out of the Barents Sea, whereas the two latter, involving protection, can take place partly outside and partly inside the region.

The West will naturally try to gather its ASW-forces near the outlets of the most probable transit routes from the Barents Sea, i.e. in the northern part of the Norwegian Sea and on the edge of the Arctic Ocean. Similarly, the SSN-forces of the Northern Fleet will try to counter this gathering of force in the same area, and will serve as 'gate-keepers' to the Barents Sea. To reduce the concentration of force in these gates, they will have to meet Western ASW-forces as far south in the Norwegian Sea and north in the Arctic Ocean as possible. In this way they will plan both the transit routes out of and within the Barents Sea and Kola by deployment in depth. The SSNs' primary use of the Barents Sea thus appears to be external transit and regional defence of Kola and the reserve area for the port-based SSBN fleet; this will probably be where the final battles over Kola will be fought. Stationary SSNs will in that case compete for available space with the SSBNs that are left. The SSNs are substantially smaller than the last generations of Soviet SSBNs which means that they have a greater operational area in the Barents Sea than their strategic counterparts. This in turn means that limitations caused by civilian economic activity become a manageable challenge for the SSN-fleet. Even if the problem of space is not urgent for the SSNs as a result of economic activity, oil production will, however, create a noise problem. To be able to discover and defend the Northern area against forward- deployed enemy submarines, SSNs require a sea as quiet as possible. Attacking submarines, like the Soviet SSBNs, will be able to hide in the noise from oil exploration or production and thereby sail unnoticed into the area. This problem is especially sensitive for the Soviet Union because Western submarines are already less noisy than similar Soviet vessels. Here noise has an opposite significance than for the SSBNs and noise thus becomes both a problem and an advantage for the Northern Fleet. Some of the problem can be solved by widespread use of active sonars which are essentially unaffected by sea noise when detecting.

Whether one wishes to consider noise as a problem or not is a matter of priorities. A certain indication of Soviet priorities can be found in the fact that the Soviet Union itself has exploration in the area, and has hinted at the possibility of cooperation in petroleum exploration and exploitation. For the Soviet military authorities, however, the noise problem involves an uncomfortable element of uncertainty. This factor will thus probably form part of their worst-case analysis.

Summary and conclusion

Two circumstances influence military operations in the Barents Sea: factors determined by nature, and civilian economic activities. The important question here has been whether the sum of these limitations reduces the Northern Fleet's possibility of achieving its assigned tasks. The natural limitations are permanent, and can hardly be changed by human influence; any scope for adaption must therefore be sought between the interest groups competing for space in the Barents Sea.

It is reasonable to believe that conflicts between civilian and military authorities on the Soviet side have remained manageable and limited till now. This is primarily because exploration activities have been fairly limited and partly outside the Northern

Fleet's operational area. It is when exploration escalates and becomes production, and also spreads geographically, that the conflicts of interests will be put to a decisive test.

From a Soviet viewpoint the problem is that they cannot control and plan activities on the Norwegian side. It is therefore possible that exploration and production areas will be located in or near the Northern Fleet's fairway. Geological structures are located where they are, and cannot be moved to suit political expediency. And since oil can be produced only where it is located in the subsea, there are obvious limitations for the scope of political activity for Norwegian authorities. For this reason, from a military strategic viewpoint, it might have been desirable for the Soviet Union to enter into a Norwegian-Soviet agreement on exploration and production, thereby obtaining a marginal possibility of influencing development. Without such an agreement, the development on the Norwegian side will take place without any Soviet 'corrective', and this could be more serious in Soviet eyes.

Some of the factors mentioned above indicate that only a limited number of SSBNs are stationed in the Barents Sea at any given time. A stationing area in the Barents Sea limits the Northern Fleet's offensive defence of the bases on Kola. By spreading the SSBN fleet out into the North Atlantic, and under the ice in the Arctic Ocean, the regional defence of Kola and the transit routes in the Barents Sea can have optimal conditions; at the same time this permits reserving space in the Barents Sea for the port-based vessels, which, conditions permitting, will be able to transit out of the region.

There are many indications that the Northern Fleet uses the Barents Sea primarily for transit, internal as well as external. The major military problem for the Northern Fleet is the channelling of transit which is mainly caused by nature. The channelling effect can be exacerbated by future production of petroleum in the transit areas, and Norway has control over one of these areas. For this reason, it is impossible to dismiss the possibility of future conflicts of interest between the two countries if and when oil development takes place.

Notes

1. The Soviet Union, along with the parties to the Svalbard Treaty, is also allowed to search for and produce oil within the four-nautical mile territorial waters of Svalbard. Evidently, the Soviet Union claims that the stipulations in the Svalbard Treaty on unimpeded access for the citizens of the treaty parties also includes the continental shelf. If this viewpoint gains ground, it will be possible to search for and produce petroleum based on the stipulations of the Svalbard Treaty. This view has been contested by Norway. Since the question has not been clarified between the two parties, it will not be treated in this discussion.
2. The Barents Sea is here limited to the area south of the straight lines drawn from the northern tip of Northeastland via the southern tip of Franz Josef's Land to the northern tip of Novaya Zemlya; it is limited by Novaya Zemlya in the east and the 500-metre contour line from the South Cape to the North Cape. The Barents Sea is then 1 196 437 square kilometres in extent. These limitations have been introduced, since this is the area in which the Norwegian Polar Institute have the most reliable data on the depth measurements. A more correct definition would have been to limit the Barents Sea either to the Egga-edge or to the Continental Shelf margin; the Barents Sea then would measure about 1 400 000 square kilometres. All percentages and calculations of area—ice depths etc.—are based on our delimitation of the Barents Sea.
3. The maps which form the basis of these calculations were prepared by researcher Torgny Vinje in 1981, and published by Norway's Geographical Survey in 1984: 'Sea Ice Distribution 1971–80'.
4. See Torgny Vinje: 'Physical Environment Western Barents Sea. Drift, composition, morphology and distribution of the sea ice fields in the Barents Sea' in *The Norwegian Polar Institute Publications* no. 179, p. 1.
5. Ibid. p. 10.
6. Ibid. pp. 15–16.
7. Ibid. p. 17.
8. Anders Elverhøy and Anders Solheim: 'The Physical Environment Western Barents Sea: sheet A: Surface Sediment Distribution', *The Norwegian Polar Institute Publications* no. 179 A.
9. Vinje, op. cit., pp. 25–7.
10. All percentages and area-calculations used in this chapter have been prepared by experts at the Norwegian Polar Institute. In this context we wish to convey our gratitude for and recognition of the

work carried out, and coordinated by researcher Anders Solheim at the Norwegian Polar Institute.

11. *The Military Balance, 1985–86*, Norwegian Edition. The Norwegian Atlantic Committee.
12. The Military Balance, 1984-85, pp. 116–21.
13. This objective has primarily been identified by Dr. Tomas Ries in his thesis on 'Northern Waters in Soviet Military Strategy', (mimeographed).
14. Willy Østreng, *Politics in High Latitude. The Svalbard Archipelago*, Montreal, McGill-Queen's University Press, 1978.
15. This point is elaborated in Jahn Otto Johansen, *Sovjetunionen og Norden. Konfrontasjon eller naboskap?* (The USSR and the Nordic Region: Confrontation or neighbourliness?), Oslo, Cappelen 1986.
16. *The Military Balance*, op. cit.
17. *Soviet Military Power*, September 1981, US Department of Defence, Washington D.C., map pp. 84–5.
18. *Soviet Military Power*, April 1984, US Department of Defence, Washington D.C., map pp. 114–15.
19. Norman Polmar, 'Sailing under the Ice', *Naval Institute Proceedings*, June 1984, pp. 121–3.
20. Willy Østreng, *Sovjet i nordlige farvann. Atomstrategien, Nordflåten og norsk sikkerhet*, Oslo, 1982, pp. 35–45. This book has been updated and translated and will be published by The Law of the Sea Institute, University of Hawaii in autumn 1986 under the title: The Soviet Union in Arctic Waters. Security Implications for the Northern Flank of NATO.
21. See Desmond Ball, 'Nuclear War at Sea', in *International Security*, vol. 10, no. 3. Winter 1985–6 pp. 3–31.
22. Polmar, op. cit., pp. 61–5.
23. John Berg, *Nærbilde av Sovjets ubåter* (Close-up Pictures of Soviet Submarines), Oslo, Sem og Stenersen A/S, 1984, p. 11.
24. Harriet Critchley, 'Polar deployment of Soviet submarines' in *International Journal*, vol. xxxix, no. 4, 1984, p. 856.
25. Ibid.
26. John Berg, op. cit.
27. Commander Reidar Skarlo, submarine inspector at Håkonsvern in Norway, has provided these estimates for the operational space needed for each vessel.
28. *Soviet Military Power*, 1986, p. 26 the map entitled, 'Modern SSBN Force Level', Washington, 1986. Military Balance, 1984–1985 , Norwegian edition, p. 116, uses slightly differing figures; it claims that the Northern Fleet has twenty-one Delta submarines and three or four Typhoons.

29. Ibid., p. 116.
30. Harriet Critchley, 'Canadian Security Policy in the Arctic. The Context for the future' (mimeographed).
31. See Edgar Ulsamer, 'Bobbing, Weaving and Fighting Smart. The US Navy looks to new systems and to new concepts to maintain its maritime edge', in *Air Force Magazine*, August 1983, p. 89.
32. Interview with State Secretary Torbjørn Frøysnes in *Stavanger Aftenblad*, 12 July 1985 under the headline: 'Poor Timing'.
33. Waldo Lyon, 'The Submarine and the Arctic Ocean', in *The Polar Record*, vol. 11, no. 75, 1963.
34. Width, length and weight are stated in most of the sources which give information on Soviet submarines. See for example various editions of *Soviet Military Power* for the years 1981–5. The height from the keel to the highest point of the tower is, however, not to be found in the same sources. We have therefore conferred with Norwegian naval experts—both civilian and military—without finding anyone able to provide exact specifications. From pictures and drawings of the boat, we have tried to calculate the height based on the figures for length and width. These calculations, which are very approximate, indicate a height of between twenty-five and thirty metres. These calculations were later presented to the same experts, who found that they gave a realistic estimate. We have therefore chosen to estimate the height to be twenty-eight metres. Similarly, the height of the Delta class is estimated to be around twenty-one metres.
35. Willy Østreng, *Polhavet i internasjonal politikk* (The Arctic Ocean in International Politics), Lysaker, Fridtjof Nansen Institute series, 1978, pp. 139–41.

4 Conclusions: Soviet options in the Barents Sea

Major alternatives

As mentioned above, Soviet decision-makers now face different possibilities in the Barents Sea area. In its simplest form, these can be seen as a choice between building up a self- sufficient offshore industry based primarily on domestic resources, or cooperative arrangements with foreign partners in one way or another. At the end of Chapter 2 we outlined the two domestic options: the rapid build-up, and the gradual learning process. These must be evaluated together with another two lines of action—industrial cooperation with Eastern Europe and with Finland. Both are closely related to the domestic options, as they all minimize the need to use hard currency for oil development in the Barents Sea.

East European equipment

Current Soviet economic plans emphasize increased trade with East European countries for obvious reasons. To the extent imports from CMEA countries can replace Western goods, precious hard currency can be saved. The USSR is running a considerable trade surplus with East European countries and is looking for valuable imports.

Obstacles to this alternative are therefore not so much economic, but technological. However, the USSR has been engaged in joint exploration for oil in the Baltic Sea with some of its East European allies, and some of the hardware has been supplied by the latter. It is very likely that these countries will be able to supply the USSR with some relevant hardware—especially ships—in the Barents Sea as well. The problem is that they seem unable to deliver the more advanced technology that is needed.

Finland

Finland occupies a special position in trade with the Soviet Union, both for economic and political reasons. This has traditionally given the country certain advantages compared to other Western countries. Trade between Finland and the USSR is conducted on a clearing basis, i.e. without money transactions between the two countries. Finnish exporters get their pay from money earned on sales of imports from the USSR. The trade is divided into several commodity groups and is thoroughly planned. In principle it must be balanced—if not from year to year, then over a five year period.

For a number of years this system has worked to the advantage of both parties. The Finns have gone out of their way to let the Soviets into their market, thereby securing their own exports to the USSR. In turn, the Soviets have obtained high quality Western goods without spending precious hard currency. This has also worked the other way around, giving Finland a strong position in the Soviet market.

Notwithstanding Finnish efforts to accept a variety of Soviet export goods, oil has for a number of years dominated imports from the USSR. In 1983 oil constituted 77 per cent of Finland's imports, and in 1984 75 per cent.[1] In terms of volume, imports peaked in 1983 at 11.5 million tons. According to the trade protocol between the countries signed in September 1984, oil exports (crude and products) were to remain at this level throughout the period 1986–90, but problems in the Soviet oil industry may make it difficult to maintain such volumes. Thus deliveries in 1985 fell to 9.9 million tons.[2] Still, Finland seems to have suffered smaller cuts in its 1985 deliveries from the USSR than some other West European importers.

All the same, this development is discomforting for the Finns. The problem is aggravated by falling oil prices. If Soviet exports to Finland go down, Finnish exports to the USSR must be reduced correspondingly. Alternatives to oil as an export item are sought for, but the Finnish market is largely saturated with Soviet industrial products. Gas imports are increasing, but will by no means replace oil in significance. In 1984 gas comprised only 3.5 per cent of total Soviet exports to Finland. The volume has been around 1 billion cubic metres in later years and is scheduled to increase to 2.2 billion cubic metres by 1990.

As a result of the negative development in Soviet exports to Finland, there seems to be no potential for increased Finnish exports to the USSR within the existing trade framework. On the contrary, Finnish exports will have to be cut to achieve balance. It is possible that the Finns will contemplate straightforward cash or credit deals with the Soviet Union, but that would leave Finland with no economic advantage compared to other Western countries. This would, if anything, be a disadvantage, since Finland is a relatively capital-poor country.

However, other factors also give Finland a special position. Economic relations between Finland and the Soviet Union are well established and stable: the two countries are used to dealing with each other. The Finnish stance in international politics, non-alignment, has also found favour in Soviet eyes. It has often been said that Finland has caused the Soviet Union fewer problems than any other neighbouring country.

Ultimately there is the question whether Finnish companies have the experience and skills necessary for large-scale offshore operations. This is not the place for such an evaluation, but it should be stressed that so far the Finns have been the main foreign supplier of hardware to the Soviet offshore sector. (For documentation see appendix 2).[3] It is difficult to imagine a large scale development in the Barents Sea without the Finns playing an important role, even if Finnish companies do not have sufficient experience in off-shore activities on their own.

The Rouble Strategy

We will consider these options (use of domestic resources, cooperation with Eastern Europe or Finland) as a whole, since they are likely to be combined in one way or another, and any combination will differ markedly from the major alternative, large-scale cooperation with Western companies. Thus in broad terms, Soviet decision-makers face two different lines of action, the rouble and the hard currency alternatives (see the bottom line in figure 1.1). The main attraction of the former is obvious. It minimizes the need to use one of the scarcest resources in the Soviet economy: hard-currency. The foreign economic situation described in Chapter 1 means that the struggle over access to hard currency will be harder than ever in the coming decade. Thus any plan for large-scale commitment of hard currency revenue will be more critically scrutinized by central Soviet economic decision-makers than ever in the last one or two decades. It will require considerable political and organizational strength at the highest level for any sector to obtain room for itself among the priority items in the tough bargaining over Soviet foreign trade in the years to come. So far the gas ministry has clearly been among the privileged industries allowed to import Western equipment on a large scale.[4] This was clearly justified by the new gas export deals concluded with Western Europe in the early 1980s. We have reason to believe that the Soviet gas industry is interested in maintaining these business relations in order to remedy some critical shortcomings in the gas production and transportation system.[5] However, as prospects for increasing gas exports to Western Europe before 1995 are rather bleak, Mingazprom may be hard pressed to convince decision-makers that it really needs much more deliveries from the West. Of course, the prospect of major new gas export deals might well alter this picture radically.

In this situation, offshore oil and the Barents Sea in particular can be seen from two opposing angles in the gas ministry: either as a diversion of scarce resources away from the gas industry proper, or as an opportunity to present a new case for a compensation project on the scale of the pipeline deals of the early 1980s—possibly much larger. We can only speculate on how this trade-off is viewed within the gas ministry and the ministry of foreign trade, but we assume that some of the characteristics of the Soviet economic system described in Chapter 1, will have a bearing on their calculations. Risk aversion explains the inertia of the Soviet energy complex and its inability to change its course of action before the problems become acute. This factor weighs heavily against any radical reallocation of resources, in particular in favour of fairly new alternatives like offshore oil. The Barents Sea option probably presents grave political and financial risks for the organizations and top-level personnel that would be involved.[6]

The lack of risk capital, a concept unknown to Soviet economic planners, points strongly in the same direction. Within the Soviet system innovation very rarely originates from below. Any alternative to existing ways of production, for example in the energy complex, must pass a certain threshold to attract the necessary investment resources. This requires both bureaucratic and political momentum and neither can be detected in favour of the Barents Sea at present. Furthermore, the threshold for a new project is probably higher now than for many years, given today's tight investment and hard currency situation.

Thus, under present circumstances, the rouble strategy stands out as the most likely course of action for Soviet decision-makers. Within this broad strategy the go-slow option, the gradual learning process, seems more probable than a radical increase in the input of resources. In practice this would mean that the Soviet offshore venture in general and in the Barents Sea will remain around the level presently planned (see table 2.2). The build-up of the drilling fleet, to which resources have already been committed, will probably be carried out. This will, however, not resolve the technological and management problems described in Chapter 2, so efficiency of operations will remain low by Western standards. This could be improved by selective purchases of Western services like seismic analyses and drilling, but again that would require hard currency.

The main drawback with the rouble strategy is that it reinforces the vicious circle outlined in Chapter 2: the more ineffective the exploration, the lower the chances of a large discovery and hence fewer resources committed to improve quality.

The hard-currency strategy

A hypothetical oil field in the Barents Sea

Since no commercial oil find has so far been made in the Barents Sea, and our geological knowledge is very limited, the following discussion must necessarily be of a hypothetical character. It seems reasonable to approach the question of oil production in the Barents Sea with experience from development of different fields in the North Sea in mind. Our purpose is here simply to reach a better understanding of the order of magnitude involved — not to present any detailed calculations for the cost of Barents Sea oil exploration.

For the purpose of this discussion, we assume that a hypothetical field in these waters, if it is to be brought into production, say, around 1995–8, will to a large extent be based on technology presently known or under development. In practical terms, this implies using one or two concrete platforms, possibly with only one shaft each. If the field is located so far from the coast as seems likely in this case, oil will probably be loaded directly to tankers. Experience from the North Sea indicates that such a platform will cost NOK 10–15 billion to construct and put into operation on the field.[7] With financial costs included, the total bill may run up to NOK 25 billion for one platform.

Such a platform could take oil from around thirty wells, each producing four to five thousand barrels per day — which brings total production to 120–150 000 barrels per day. It would probably take a couple of months to drill each well, which implies that it might take three to five years to reach maximum production. The operating costs of such a platform, including personnel, supply, transportation, maintenance etc., will probably be on the level of NOK 450–500 million annually.

In this hypothetical example, we assume that the Soviets purchase such a platform (or if necessary, possibly two platforms), and simultaneously enter into a management or technical assistance arrangement with Western companies which would make them responsible, in deed if not in words, for the production from the start and for a period of, for example, five years. These would be used for on the job training of Soviet personnel. The price of such an 'assistance' contract would probably be fairly close to normal operating costs.

Furthermore, we presuppose that the Soviets would demand 100 per cent credit, at least for the capital equipment - to be repaid with hard currency proceeds from future oil production. As for the interest rate, the Soviet standard in dealing with the West is very low, 7.5 per cent. However, the difference between this level, which the Soviets may insist on, and the commercial rate, would simply be added to the purchasing price. Seen from the Soviet angle this is just a matter of bookkeeping, but it might create difficulties for Western companies involved. A crucial factor for the latter is their relationship with credit institutions.

A very important question is how long it would take the Soviets to repay the capital investments in a hypothetical Barents Sea field. We can make a comparison with the gas pipeline deals, for which Western experts estimate that the Soviet import bill for pipe and equipment amounts to $15–20 billion (from 1981 to 1984), which will be repaid over ten years from 1985 to 1994. If financial costs are included, the annual repayments are probably in the order of $1.6-2.2 billion over this period—which is easily covered by the revenues from gas sales to Western Europe.[8]

If one Barents Sea platform were to be repaid according to a similar pattern, Soviet hard currency expenditure should amount to NOK 2–2.5 billion per year (including interest payments). If operating costs are added, the annual bill would run from NOK 2, 5–3 billion or approximately $350–430 million (at the present exchange rate $1 = NOK7).

If the capital investments were to be repaid over a shorter period, which is common in the North Sea, the outlays would naturally increase. Adding another platform, which might be necessary under certain reservoir conditions, largely doubles all the costs. Capital outlays (including financial costs) might then run as high as $7 billion and annual operating costs up to $140 billion. If repaid over only five years, Soviet hard currency expenditure might in the case of two platforms amount to approximately $1.5 billion annually. In other words, there is a very strong element of uncertainty in all such calculations, as nobody knows what the hypothetical field would look like. We can, however, presume that Soviet hard currency outlays on a yearly basis might range from $350 million at the lower end to around $1.5 billion at the higher. Any of these numbers are of considerable importance to Soviet foreign trade, especially with Western Europe: total Soviet imports from this region stabilized at the level of $12-14 billion in 1982-5; imports of machinery, equipment and pipe from Western Europe were around $6 billion in 1984.[9] In comparison, total Soviet imports from Western Germany, its largest trading partner in the West, amounted to $3.9 billion in 1985.[10]

The figures above can also be evaluated in relation to the debt situation of the USSR. Total Soviet hard currency debt at the beginning of 1986 amounted to $9.6 billion,[11] compared with a possible credit for one or two offshore platforms of $3–4.3 billion. Still, it must be stressed that the USSR has very limited foreign debt relative to the size of its economy.

In spite of the uncertainty underlying any calculation of this kind, we draw the following conclusion: if the Soviets choose to develop an oilfield in the Barents Sea primarily using Western equipment and management (the latter concealed as technical assistance), such a deal would involve hard currency expenditure and indebtedness of considerable economic and political importance to the Soviet leadership. The development of any one field would have a noticeable impact on Soviet trade with the

West. If several fields are envisaged, the level of economic co-operation might soon approach, possibly even surpass, that of the gas pipeline deals of the early 1980s.

These are the prospects facing Soviet decision-makers in this area. Below, we try to evaluate the trade-offs they are making in this regard. In addition to the factors mentioned there, the Soviet organizations will obviously ask themselves how much oil they will need to produce, how fast and at what price, to repay their outlays in hard currency. Again, accurate predictions are meaningless, but experience from the North Sea may give some clues: for example, it is presently estimated that the capital investments of the Statfjord C platform will be repaid in two and a half years (with an oil price of $15 per barrel). This field has exeptionally favourable production characteristics, however. In the case of a smaller field like Ula, it takes eight years to cover the investments (NOK 8–9 billion) at a $15 per barrel oil price, but only three years if the price goes up to $20.[12] Such calculations are obviously of great interest to Soviet decision-makers as the figures indicate that an oil field in the Barents Sea, whether large or medium size, has the potential to become a source of net hard currency income within a limited number of years—if the Soviets are willing to take the risks involved (see below).

The form of cooperation

A common cooperative arrangement between the Soviet Union and Western companies delivering production equipment is the compensation deal, with payment in kind after production has started. This reduces Soviet capital and hard currency expenditure to a minimum. It makes the foreign investor responsible for establishing the plants and for bringing them into production. With regard to the Barents Sea, such an arrangement would solve two critical problems on the Soviet side: investment capital and management. If Western companies could be persuaded to take the risk of petroleum exploration and bring in their offshore experience, this could with one stroke eliminate the bottlenecks in the Barents Sea venture. So, on paper at least, the hard currency strategy, or perhaps one should call it a 'Chinese model', may seem very attractive. However, within the Soviet system the obstacles to it are formidable.

First, such a scheme would be extremely sensitive in political terms. It would underline Soviet technological backwardness and dependence on the West in a very uncomfortable manner. As the political fate of a compensation project would be very hard to predict, it is a highly risky course of action for the organizations involved, primarily Mingazprom and the ministry of foreign trade.

Second, the project is equally uncertain in foreign policy terms. It immediately raises the question: can the West be trusted as trading partners after the pipeline sanctions? We do not know what lessons the Soviets have learnt from that episode, but it has undoubtedly underlined the basic political character of such large-scale deals. The Soviets probably realize that American technology or components can hardly be avoided, even if reduced to a minimum, in a Barents Sea compensation project. On the other hand, Soviet foreign policy makers might regard it as a new opportunity to create tension over East–West trade within the Western alliance.

At any rate, such a scheme must be evaluated in the larger and highly volatile context of superpower politics. Again, this is not an attractive environment for the Soviet organizations that might be interested in a large-scale development of the Barents Sea. The third obstacle lies in the sensitive military situation in the northern waters. Our analysis in Chapter 3 indicates that the Soviet military has several reasons for concern about future large-scale Western oil activity in the Barents Sea:

- The physical installations would limit the manoeuvring space available for the surface vessels of the Northern Fleet which is already a critical point. Since they can hardly hope to stop oil development on the Norwegian side, the Soviet military may be even more sceptical about similar activities in their own sector.
- Increasing 'background noise' as a result of oil exploration would make it easier for foreign attack submarines to operate unnoticed in the Soviet Barents Sea. On the other hand, the same factor would also provide better protection for Soviet strategic submarines. How military commanders evaluate this trade-off is very hard to say, but we suspect that they would prefer to be without the increasing insecurity which follows from the background noise.
- In addition to these objective factors, the Soviet military are probably worried about these prospects for psychological

reasons. They would resent a Western presence in an area so far dominated by Soviet—mainly military—activity. As indicated in Chapter 3, the Soviets have always been extremely suspicious of Western civilian activities in the Arctic, for example on Svalbard, and there is no reason why this suspicion should not be even stronger the closer one gets to the core of the Northern Fleet. The forward-based naval strategy promoted by the Reagan administration obviously adds to these concerns. As mentioned above, it is already difficult for the Northern Fleet to fulfil all its tasks, especially protection of the stationing area, the transit lines and the base, because of geographical factors. These problems would be exacerbated by Western petroleum activities. The Soviet military might come to feel that the West is moving rapidly northwards with both submarines and oil rigs.

Consequently, we conclude that the Soviet military would have preferred to exclude petroleum exploration from the ice-free waters in the Barents Sea. If this proves impossible, they would like to limit it as much as possible, in particular to ensure that the major transit lines out of the Kola base are not negatively affected. In the ice-infested eastern areas, the Soviet military would probably have fewer reservations about oil exploration as such, but Western participation in any part of the Barents Sea is undoubtedly contrary to their interests. This is particularly relevant for a large-scale compensation deal, since it would imply that some of the control over future activity in a highly sensitive area would be transferred to Western companies. The Soviet military probably realize that such an arrangement would imply that production would follow very rapidly after exploration, if it leads to a major discovery.

Under these circumstances, there is every reason to ask if any form of large-scale Western participation in Barents Sea oil exploration and development is at all conceivable. We will approach this question by examining the conditions which must be fulfilled for the hard currency strategy to become a real option and not only part of an academic discussion. As indicated above, this is primarily a political decision which can be taken only at the highest level, eventually within the top party leadership. In other words, this strategy is based on the premise that the central decision-makers make a deliberate choice in favour of (1) the

Barents Sea as part of a long-term solution to Soviet energy and foreign trade problems, and (2) the use of Western technology and capital for this purpose in an arrangement acceptable to both parties. Such a decision could be made in the near future if current oil production difficulties continue and the leadership is forced to look for new long-term solutions instead of prescribing more of the old medicine. We will refrain from any assessment of the likelihood of such a decision being made. It will also depend on how the leadership looks at the current foreign trade situation, which again is influenced by its foreign policy in general. There are simply too many uncertainties to make any meaningful prediction. Nevertheless, such a move by the party leadership is certainly not inconceivable.

If we then proceed to assume that such a decision in favour of Barents Sea development with large-scale Western participation is taken, what of the effects? First of all, the most important political uncertainties would be removed from the operational organizations. The party leaders would take over those risks, thereby giving the go-ahead for managers in the bureaucracies involved. The gas industry could go ahead with detailed plans for development with assurance of necessary resources. The ministry of foreign trade could enter into negotiations with Western companies. More concretely, we could expect to see the impact of a top-level decision in the following ways:

- Re-organization: The offshore industry in general and the Barents Sea would be given a higher status in the relevant organizations, especially Mingazprom. (It is also possible that offshore oil would be taken over by the oil ministry). New bureaucratic bodies or standing committees would be set up to integrate the activity in various ministries.[13]
- In its foreign policy the Soviet Union would seek some kind of political reassurance from key Western governments, as the Soviets did over the gas deals in the late 1970s and early 1980s.
- Negotiations on concrete terms of cooperation would be taken up by the ministry of foreign trade with several Western companies or combinations of them. The Soviets would show the flexibility necessary to obtain enough interest on the other side, always maintaining as much competition among Western companies as possible. Access to other raw materials might be used as an extra carrot.

- Some hard currency, considerably more than has been the case so far, would be used as 'seed money' for the large projects. The purpose would be both to whet the customer's appetite and to speed up exploration efforts by way of key services from the West, for example re-analyses of seismic data.[14]

The top-level decision we have in mind clearly means that the military objections would be overruled. Again all we can say is that this is within the range of the possible. The Soviet leadership under M. Gorbachëv seems to be more intent on keeping the military under party control than has been the case lately.[15] It is not inconceivable that the party leadership will give priority to economic concerns despite military objections. If the domestic economic situation becomes even more difficult, and oil production falls dramatically, such a move would become more likely than it is today. (The outcome of the *Marlene Østervold* incident can also be taken to indicate that civilian authorities may prevail over military intransigence.) The question remains however, whether the military could still silently obstruct co-operation with Western partners in the Barents Sea in practice, even after a top-level decision were taken. Nevertheless, it is a fact that military reservations have not been strong enough to prevent oil exploration. (But it is possible that the first exploration activity was perceived more as a natural extension of a well established Soviet research tradition in the Arctic than as a prelude to industrial utilization of the shelf.) However, the conflict between civilian and military interests has not yet been put to a critical test.

This speculative argument could be taken even two steps further: what concrete terms would the Soviets have to offer to attract Western business interests to the Barents Sea, and what kind of Western responses could they expect?

The anticipated framework for a deal of this type is the compensation project with payment in kind, with only very limited entrepreneur or service contracts to be paid in cash. In other words, Western companies would be invited to participate in the exploration at their own cost, possibly in joint-ventures where the foreign companies provide most of the capital, technology and management and the Soviets provide some hardware and personnel. The Soviet organizations would certainly use such an opportunity to train their own field operators and manpower in general.[16] In return, the Soviets would have to offer very attractive terms for repayment in kind. If Western companies are to take all or most of the risk, the conditions concerning crude sharing must be very favourable. Furthermore, if Western business firms are to carry the risk, they would ask for some kind of control over future operations in the area, so that they can themselves evaluate the risks and the costs. By pressing for risk sharing, the Soviets are in effect preparing for a more integrated and committing Western participation than if Western companies were used only as technical operators, receiving their payment in cash. The Western response to proposals along these lines would be influenced by both commercial and political considerations. The attitude of the companies involved would depend on several factors. These include perceptions of the world oil price for the late 1990s[17]; present financial position of the company—in particular its access to crude oil in the long term; other interests in the Barents Sea area, in particular raw materials or the mining industry on the Kola peninsula; other interests in trading with the Soviet Union; and relations with the Norwegian government.[18] Today's over-capacity in the Western offshore industry might induce some companies to make favourable offers to the Soviets concerning, for example, the use of exploration platforms.

On the political side, reactions in Western Europe will most likely be very positive. The Barents Sea could be seen as a natural continuation of *détente*, pursued for example in gas export deals. Such attitudes are likely to be expressed in various ways in the major West European countries, reflecting different traditions in government to business relations.[19] Despite differences in form, there is no reason to doubt that all the major West European governments would give the Soviet Union reassuring political support in favour of large-scale industrial cooperation in the Barents Sea. The governments would probably be only too eager to obtain as large a share as possible of the contracts involved. The Cocom rules seem unlikely to create any serious difficulties for such deals.

The American position is much more difficult to predict. It is still unclear what lessons various parts of the administration have drawn from the pipeline controversy: does the outcome signal that an embargo line without unanimous allied support has been abandoned for good? Or is the opposite interpretation more common: the pipeline embargo has demonstrated that such policy instruments can be effective in forging western consensus on East–West relations and may therefore very well be used again when required by a particular situation. (Our purpose is not to predict American reactions, but rather to evaluate Soviet options faced with this uncertainty.)

Given these possible Western responses, which kind of partnership would the Soviets prefer? On paper there are three major options: a Nordic, a Western European and an all-Western, including the United States. (Other partners, particularly Japan, can of course not be ruled out.) In practice these can naturally be combined in many different ways, but for the sake of the discussion, we will look at them separately from the Soviet point of view.

Cooperation with Nordic companies would have several advantages for the Soviets: they are used to dealing particularly with Finnish, and to some extent Swedish companies in the development of the Northern regions. The Barents Sea could be seen simply as an extension of such cooperative ventures. In addition, the Soviets seem to feel politically comfortable in business relations with its small Nordic neighbours. There is also a tradition in Soviet policy to look at the Northern areas in such a regional perspective, thereby conveniently excluding the larger Western powers. The major drawback of this option is that the Nordic companies likely to be involved can hardly supply all the inputs necessary for large-scale development of the Barents Sea, at least not without important sub-components from non-Nordic companies.

The Western European option would as mentioned above, be a natural continuation of the Soviet import policy during the gas negotiations in the early 1980s. There is reason to believe that the Soviets would be interested in using Barents Sea contracts in order to improve their standing in important political and economic circles in Central Europe which is vital for future gas deals. (If the Soviets are looking seriously at the British gas market, this also applies to their trade with the United Kingdom). The Soviets have well established business contacts with important equipment manufacturers in Germany, Italy and France. The Soviet gas industry has cooperated with such companies now for a number of years, and we know that the Soviets prefer to maintain such old ties. In addition, the Western European option opens up a more competitive situation than an exclusively Nordic approach would. The West European option is also politically attractive for the Soviets since they could expect to reach a clear political understanding with major Western governments as a basis for the business deals. These governments would carry so much political weight, as the pipeline dispute demonstrated, that the Soviets can be rather certain that whatever happens in trans-Atlantic relations, Western countries will adhere to their contracts with the Soviets, in particular in such large-scale deals. In addition, the Soviets might see a possible foreign policy advantage in creating a future parallel to the pipeline controversy over the Barents Sea, should the American government advocate a new embargo.

Then there is the all-Western option, including American companies. It is not likely that West European companies could deliver projects of the kind discussed here without components produced under licence from American companies. So from a Soviet point of view the danger of an embargo would always remain. This might encourage Soviet decision-makers to include some American business interests in a compensation project which would also increase the number of competitors on the Western side.

In our view, the most likely combination of these options would be considerable participation from both the Nordic region and Western Europe to assure a competitive situation and to build broad political support on the Western side for such deals. In addition, the Soviets might like to include a minor American share, possibly as a business hostage against embargo initiatives from Washington. However, the Soviets probably realize that this would not make much difference to American decision-makers. So, if they were to choose to include American companies, the motivation would be primarily commercial.

The analysis in this chapter leads us to conclude that large-scale Western participation is a prerequisite for major progress in oil exploration in the Soviet Barents Sea. If the Soviets really want to make this area a serious option for oil production in the 1990s, extensive outside assistance will be needed in both exploration and production. The major question is whether they are prepared to pay the price for such participation—in terms of hard currency expenditure and political and military risks. We have to leave this question unanswered, as we do not know the priorities of the Soviet decision-makers involved. We can only point to the trade-offs involved.

In concluding we will, however, underline that the fall in the world oil price will not affect the Soviets in a way similar to Western companies. On the contrary, a sharp price decline makes it more important than ever to increase, or at least to maintain, the present level of exports. As we have shown above, the Soviets have very limited freedom of action in their foreign trade. So the prospects of a falling, or stagnant, real price of oil would most likely encourage Soviet decision-makers to look for new areas of potential production. In this evaluation, the Barents Sea remains a promising, but risky and uncertain card—yet to be played.

Notes

1. *Vnesnyaya Torgovlya SSSR*, 1984.
2. Official Finnish statistics.
3. The Finns have also formed a joint 'Arctic Technology Committee' with the Soviets, consisting of Finnish companies, trade and research institutions and from the Soviet side the State Committee on Science and Technology (GKNT), the gas ministry, and some research institutions. One of the expressed Finnish aims is to influence Soviet choice of technology for Arctic offshore operations.
4. It is difficult to estimate the exact level of imports to the gas industry. According to Ed Hewett, 'Near Term Prospects for the Soviet Natural Gas Industry and the Implications for Foreign Trade' in *Soviet Economy in the 1980s: Problems and Prospects*, Joint Economic Committee, U.S. Congress, Washington, 1983, the total value of pipeline equipment imported from the West in 1981–4 was $15–20 billion.
5. This includes *inter alia* large-diameter pipe, compressors and sulphur treatment equipment. The USSR is not unable to produce such items, but foreign equipment is more efficient and Soviet equipment is not always available in time or in the necessary quantities.
6. Indeed, one of a number of interpretations of the fate of deputy trade minister Shuskov, who was arrested for smuggling in January 1986, was that he had lost his political fortune due to his central role and responsibility in the unsuccessful Sakhalin project (see Chapter 1).
7. The Statfjord C platform is, by comparison, estimated to cost approximately NOK10 billion.

8. Ed Hewett, op. cit. pp. 409–10.
9. *Vneshnyaya Torgovlya SSSR, 1984* and *Vneshnyaya Torgovlya*, 3, 1986.
10. *Vneshnyaya Torgovlya*, 3, 1986.
11. *PlanEcon Report*, 17–18, 1986.
12. Estimates made by the Institute of Industrial Economics in Bergen—and communicated directly to the Fridtjof Nansen Institute for which we are very grateful.
13. The recently established bureau for energy policy attached to the council of ministers and headed by Boris Shcherbina is an interesting attempt to coordinate the fragmented ministry structure in the energy sector. Such a body could conceivably also be of relevance for large-scale decisions concerning the Barents Sea. According to the *Financial Times*, 3 May, 1986, '[the bureau] also has power to switch between sectors the vast investment funds and resources which the Soviet Union has poured into energy'. Time will show if this is a reality.
14. According to a recent CIA report the Soviets need western technology and equipment for geophysical surveys and drilling operations *inter alia* in the Barents Sea. *Soviet Needs for Western Petroleum Technology and Equipment*, Central Intelligence Agency, Washington D.C., 1986.
15. The most prominent indication of this may be their weakened representation in the politbureau.
16. In addition, the presence of Soviet personnel on Western installations may reduce the fear of military abuse of oil equipment.
17. The relevance of the world market price will to a large extent depend on the particular terms of a compensation contract. If payment in oil is indexed according to the world market price, price expectations will have little significance. The other extreme, payment in fixed volumes, makes price expectations a paramount factor. Eventually, a contract may include elements of both principles, probably more of the former principle than of the latter.
18. It has been indicated by Norwegian authorities that they will not tolerate operations by Western companies in the Soviet Barents Sea if these go counter to Norwegian political interests. The implication is that such actions would have negative repercussions on the companies' activities on the Norwegian shelf. This view was expressed by then state secretary Nils Morten Udgaard at a meeting of the Norwegian Petroleum Society, 21 Octboer, 1985.
19. In countries like Italy and France, governments are heavily involved in the industry and are likely to take an active part in any cooperative project of this size with the Soviets. In the Federal Republic of Germany, the ties between government and industry

are of a more discrete nature, and any expression of interest from the German side will come from private companies but with a general backing from the government. In the case of the United Kingdom, the government has recently signed an energy cooperation agreement with the USSR—which will serve as a framework for new offshore contracts. The British government has already expressed its support for such commercial deals (see *The Financial Times*, 25 April, 1986).

Appendix 1
The organization of the Barents Sea offshore research, planning, exploration and development: some notes and comments on Figure 2.2

The organizational chart (figure 2.2, Chapter 2,) indicates the institutions that, to our knowledge, participate in decision-making, planning, research, education, exploration and development in connection with the offshore activity in the Soviet Barents Sea. The chart is certainly not complete and is to some extent based on guesswork. We find, however, that the information compiled represents new knowledge about the Barents Sea organization and about the Soviet offshore industry in general. Index letters refer to the chart.

A: KPSS—The Communist Party of the Soviet Union

Nothing is known about the central party organs' participation in decision-making on the Barents Sea. It is, however, expected that a substantial Western involvement in offshore exploration and development would have to be approved by the Politburo. The issues decided by the Politburo are normally prepared by the state institutions and the Central Committee department concerned. A formal proposal must first be passed by the Secretariat of the Central Committee, which is led by the secretary general. At present the secretary responsible for energy policy is V. I. Dolgikh, who is also a candidate member of the Poltiburo. It is likely that he would present any case on Barents Sea offshore development in the Politburo.

If the matter were to include heavy involvement by Western companies, it might be presented by someone else, and would in any case be of interest to several of the Politburo members—inasmuch as it concerns defence (Zaykov, Sokolov, Chebrikov), foreign policy (Gorbachëv, Shevardnadze, Gromyko) and economy (Ryzhkov, Talyzin) etc.

A-1: The Heavy Industry and Power Engineering department of the Central Committee of KPSS[1]

As mentioned above, this department is supervised by V. I. Dolgikh (born 1924). I. P. Yastrebov (born 1911) has been its chief since March 1984, when he took over from Dolgikh. He had then held the position of first deputy department chief for twenty-two years. In 1983 the department changed its name from 'Heavy Industry' to 'Heavy Industry and Power Engineering' which illustrates the growing importance attributed to energy policy among the top Soviet policymakers.

B: SOVMIN SSSR—The Council of Ministers of the USSR

Formally the SOVMIN is the equivalent of the government in a Western country. However, responsibility for overall policy-making belongs to the party, and the principal function of the SOVMIN is implementation. But it is difficult to draw a strict line between the functions of SOVMIN and those of KPSS, the most prominent reason being the personnel overlapping between the state and the party, which plays a central role at all levels of Soviet politics: many officials occupy important positions in the government administration as well as in the party.

Traditionally the SOVMIN is first and foremost engaged in economic policy and administration. The Council of Ministers consists of more than a hundred ministers and state committee chairmen. Naturally, it cannot function as a decision-making body. The central role of the council is played by the Presidium, which consists of the SOVMIN chairman (sometimes referred to as the prime minister) and a dozen or so deputy chairmen. In many fields these deputies seem to duplicate the secretaries of the Central Committee. Dolgikh's opposite number is Boris Shcherbina (born 1919, appointed 1984); Shcherbina is, however, believed to be more narrowly specialized on energy policy.[2] On behalf of the chairman, N.I.Ryzhkov, who is formally in charge, Shcherbina is responsible for the day-to-day control of all ministries and state institutions in the energy sector.

Both the Central Committee Secretariat and the SOVMIN Presidium play a coordinating role in the different policy fields. In the energy sector for instance, there is an abundance of ministries, state committees and other organs. The need for coordination is even greater when energy interests are confronted with those of other sectors. As the Soviet economy grows more complex, the coordination problem becomes more and more accute. In agriculture an attempt to solve the problem was made by merging several ministries into one superministry.[3] In the machine building and energy sectors the Soviets seem to have chosen another strategy—preserving the original organizational structure but adding a new coordination bureau on top of it, a kind of enlarged secretariat for the deputy SOVMIN chairman. As yet we do not know the powers of these new institutions.[4,5]

B-1: Energy policy coordination bureau of the USSR Council of Ministers

Probably established in early 1986. Headed by deputy SOVMIN chairman Boris Shcherbina (see B).[5]

A/B-1: Local authorities, Murmansk[6]

Regional and local party and state authorities in Murmansk are taking part in the offshore development process. Like the central institutions they are to control and coordinate the activities of the operational agencies. In addition one can expect that regional and local officials are furthering the interests of Murmansk in the central decision-making apparatus in Moscow in order to attract resources to the area.

C: MINGAZPROM SSSR (MGP)—Ministry of the Gas Industry of the USSR (A-U = All-Union)

In 1978 this ministry took over the responsibility for offshore exploration, development and production. It is now playing the central role in offshore R & D, planning and exploration in the Barents Sea and is also important in the contacts with foreign offshore companies. It is, however, evident that MGP is primarily implementing the general policy decisions made at a higher level, probably in the Central Committee apparatus and the Council of Ministers; may be even in the Politburo. (See A, A-1, B, B-1).

Central MGP personnel

The new minister, V. S. Chernomyrdin (born 1938, appointed 1985), recently elected full member of the KPSS Central Committee, is taking an active interest in offshore matters. He has met with Western offshore companies several times and visited Norway in July 1985.[7] With his relatively young age and fast career he may be considered to represent the new generation of more dynamic and technocratic managers under Gorbachëv.

First deputy minister, V. I. Timonin, is said to have special responsibilities for the offshore industry. In 1983 he was recruited from the oil ministry. He had for several years been chief of an important oil production association in the Kazakh republic, Mangyshlakneft.[8]

Deputy minister, N. M. Nemchinov (born 1928, appointed 1979)[9] is probably responsible for the mobile drilling rigs and the ministry's fleet of support vessels.[10] We have reason to believe that he has been directing the establishment of the Barents Sea exploration agency, Trest Arktikmorneftegazrazvedka (C-35) in Murmansk.

Another deputy minister, S. R. Derezhov, (born 1932, appointed 1977)[11] does also have offshore responsibilities, probably for the stationary production installations. If this is so, he has not been directly involved in the Barents Sea operations so far, but will be in case of a field development.[12]

C-1: The Scientific-Technological Council of the MINGAZPROM

The council is directing the offshore research and development work of MGP organizations and also their cooperation in this field with, among others, Goskomgidromet (D), AN SSSR (E), Minvuz SSSR (I) and Gosstroy SSSR (J). At present much attention is given to natural and climatic conditions for oil and gas exploration and development in the Arctic; the impact of the petroleum industy on the environment; technological solutions etc. Deputy Chairman of the council, R. S.Vorobeva, seems to be responsible for offshore research and development information from the council.[13]

C-2: The geological administration of the MINGAZPROM

We believe that this administration, which belongs to the central ministerial structure, is responsible for collecting and analyzing seismic data from the Barents Sea. This task is implemented by VMNPO Soyuzmorgeo(C-21) and NII Morgeofizika (C-22) through a permanent geological-geophysical expedition in Murmansk. This branch of the MINGAZPROM will naturally have a strong say over where to spud wildcats and eventually over the allocation of exploration drilling rigs between the different offshore areas.[14]

C-21: VMNPO SOYUZMORGEO, Riga—Scientific production association for offshore geology and geophysics (A-U)

Soyuzmorgeo is probably situated in Riga, where there are several maritime research, design and production organizations. Having participated in establishing the permanent geological-geophysical expedition (C-23) in Murmansk, Soyuzmorgeo seems to play the central operational role in collecting the seismic data. Its deputy general director, B. K. Ostisty, is stationed in Murmansk, and is simultanously deputy director of NII Morgeofizika, which also is part of the expedition. Ostisty is believed to be head of the seismic operations conducted by the Mingazprom from Murmansk.[15]

C-22: NII MORGEOFIZIKA—Offshore geophysical research institute

The institute is probably subordinated to Soyuzmorgeo (C-21). Its task in Murmansk is to process the seismic data received from surveys in the Barents Sea. The institute is believed to use relatively modern and powerful computer hardware, while advanced software of Western standard may be nonexistant in Murmansk.[16, 17]

C-23: The permanent geological-geophysical expedition, Murmansk

We do not know whether any other organizations, in addition to Soyuzmorgeo and Morgeofizika, take part in the expedition. As an organizational form the 'permanent expedition' is not uncommon in the

USSR, especially not when it comes to activities like geology and geophysics. If commercial oil fields are discovered in the Barents Sea, one may expect the expedition to be reorganized into a more independent unit[18] (see C-2, C-21, C-22). The expedition is believed to cooperate closely with the agency conducting the exploration drilling, Trest Arktikmorneftegazrazvedka (C-35).

C-3: GLAVMORNEFTEGAZ—Offshore Exploration, Development and Production Main Administration of the MINGAZPROM

This main administration (or directorate) belongs to the central ministerial apparatus and is at the same time directing the activities of the subordinated operational exploration and production units in all the offshore areas of the USSR. Trest Arktikmorneftegazrazvedka (C-35), Murmansk, is believed to be directly subordinated to Glavmorneftegaz.

It is probable that this organization has some power to influence allocation of resources between the different offshore areas, i.e. both through the planning process and within the framework of the plans. As shown in Chapter 1 the general tendency in the Soviet economy is to give priority to short-term plan fulfilment. This is probably true also in the case of Glavmorneftegaz which may imply that priority is given to the areas where production already is going on, like The Caspian Sea. Glavmorneftegaz fulfilled its production plan for 1985, thanks to oil from new fields in the Caspian Sea and off Sakhalin. Exploratory offshore drilling seems to have fallen substantially behind the plan (see C-2).[19]

C-31: VNIPI MORNEFTEGAZ, Moscow—Offshore Exploration, Development and Production Research and Design Institute of the MINGAZPROM (A-U)

VNIPIMorneftegaz is the umbrella research and design institute for the offshore branch of the gas industry. Its field of interest seems to coincide with that of Glavmorneftegaz (C-3); it is therefore tempting to conclude that the institute is subordinated to Glavmorneftegaz.[20] There is also reason to believe that several regional offshore research and design institutes are subordinated to VNIPIMorneftegaz.[21] GosNIPI Gipromorneftegaz (C-33) and Spetsmorneftegazproekt (C-34) probably belong to this group.

The VNIPIMorneftegaz institute covers a broad range of topics. Economic aspects of offshore activity and equipment are studied regularly.[22] The institute is frequently in touch with foreign offshore companies. It evidently plays a central role in assessing foreign offshore technology and in deciding whether to import it or use domestic products.

It is probably in the interest of the research and design institutes to encourage foreign technology imports, while at the same time maintaining a certain level of domestic production. In the Soviet press the tendency to prefer Western technology has been criticized[23] (certainly not without reason). But to the experts it must be more comfortable and safe to use Western technology in the risky parts of a project.

NTTs—The Scientific-Technological Centre

Attached to the institute is the Scientific-Technological Centre which is coordinating the R & D activity in the offshore sector in accordance with the directions given by Mingazprom through its Scientific-Technological Council (C-1). The Centre has a Council of Directors which consists of the leaders of all offshore related research and/or design institutes within the Glavmorneftegaz field of responsibility.[24] At the meetings of the Council of Directors, representatives of related institutions may participate. The GKNT (K) is probably always present. Goskomgidromet (D), Academy of Sciences (E), Minvuz (I) and Gosstroy (J) institutions are also attending frequently.

The director of VNIPIMorneftegaz, P. A. Maksutov, is also leader of the NTTs. The Centre has several 'Problem Councils' working to solve different tasks put forward by the ministry in the Centre's working programme. The 1983 programme included research and design work on such things as modules for platform superstructures, grounded icebergs for exploration drilling, control systems for drilling, automatic EDP-based control systems for oil and gas production, fixed drilling and production rigs.[25]

In November 1984 the journal of the Mingazprom published a review of the findings of an audit of VNIPIMorneftegaz undertaken by the ministry's central control and audit department. It revealed several rather grave irregularities in the administration of the organization. The director and four deputy directors were reprimanded[26] but are still in office. A thorough audit of any Soviet organization will invariably reveal irregularities. It is worth noticing, however, that another key Barents Sea offshore organization, Arktikmorneftegazrazvedka (C-35), has been the object of similar scrutiny—with even more alarming results.[27]

C-32: VMNPO SOYUZMORINZHGEOGIYA, Riga—The scientific production association for marine geology and engineering (of the MINGAZPROM) (A-U)

This association has been responsible for working out new offshore construction norms (VSN 51.2-84/Mingazprom), which were adopted by the ministry and made effective from 1 April 1985. The norms have been agreed with Gosstroy (J), the State Committee for Technical

Inspection (Gostekhnadzor SSSR), the health ministry (Minzdrav SSSR), the ministry of fisheries (Minrybkhoz SSSR) and the ministry of land reclamation and water resources (Minvodkhoz SSSR). The norms are based on hydrographic, geodesic, geological and hydrometeorological studies of the continental shelf in order to design offshore installations, and also to decide where to locate mobile drilling rigs. The norms were worked out with assistance from GosNIPI Gipromorneftegaz (C-33) and VNIPIMorneftegaz (C-31).[28]

SITs—morneftegaza—Specialized information centre for marine geology, geophysics and development of offshore oil and gas fields

The information centre is established within the Soyuzmorinzhgeologica organization. It functions as a data base and an information distribution system serving offshore related subscribers on a national basis. It consists of two data banks:

(1) The 'shelf' comprises a wide range of scientific, technological and industrial information about geological research on the ocean seabed and the development of its mineral resources. The 'shelf' data bank came into operation in 1983.
(2) The 'parus' gives information on offshore technology in a broad sense, including foreign technology and experience. ('Parus-1' has been operating since 1981; 'parus 2' is a modified version.)

The data banks have been developed by the all-union research institute VNII Morgeo (G-1), which belongs to the ministry of geology (G), in cooperation with the scientific production association NPO Parma, both probably situated in Riga, Latvia.[29]

C-33: GosNIPI 'GIPROMORNEFTEGAZ'—State Offshore Research and Design Institute, Baku

The institute is responsible for working out the initial specifications to produce simulators (*trenazhery*) and other teaching materials for offshore enterprises.[30] It is engaged in working out norms of hydrometeorological conditions in order to design offshore installations.[31] As mentioned above, Gipromorneftegaz participated in preparing the new offshore construction norms (C-32).

C-34: SPETSMORNEFTEGAZPROEKT, Simferopol—Special design institute for production and design

This institute is probably subordinated to VNIPIMorneftegaz (C-31). Traditionally it has been specializing in pipeline projects, but is now engaged in the whole range of offshore projects. Spetsmorneftegazproekt is the head design institute for offshore development in the Black, Azov and Baltic Seas, it is also operating in Romania, Poland

and Vietnam.[32] Some of its work may be of interest to the Barents Sea development too, for example the studies on modification of offshore rigs for ice conditions,[33] and on organizational integration within the offshore sector.[34]

C-35: Trest ARKTIKMORNEFTEGAZRAZVEDKA, Murmansk—The Arctic Offshore Exploration Trust

Arktikmorneftegazrazvedka is probably directly subordinated to Glavmorneftegaz (C-3). The name 'trest' (or 'trust') indicates, however, that it is an industrial unit on a lower level. This may argue in favour of the existence of an intermediate institution between the trust and the Glavmorneftegaz but this is not very likely. If commercial oil finds are made, the trust will most probably be turned into an all-union production association (VPO Arktikmorneftegazprom) like those operating in other offshore areas. The trust consists of, among others, the following operational administration units: (1) offshore drilling (2) construction and technical transportation (3) special equipment[35] and (4) Arktikmorneftegazflot (C-36). See Chapter 2 for further information about this organization and its activities.

C-36: ARKTIKMORNEFTEGAZFLOT—The Arctic Offshore Fleet Administration of Trest Arktikmorneftegazrazvedka (C-35)

This administration runs the specialized offshore fleet stationed in Murmansk consisting of drilling rigs, supply, accommodation and other service vessels (see Appendix 2). It is uncertain whether the geological and seismic research vessels based in Murmansk are also run by this organization.[36]

C-36: The professional-technical school, belonging to Arktikmorneftegazrazvedka, Murmansk

The school educates workers not only for the needs of Arktikmorneftegazrazvedka, but also for other MGP organizations. In a speech in February 1984 the then gas minister Dinkov praised its founder, director Glukhov, for having turned the school into a leading educational institution.[37] The school has its own deep-diving programme, teaching underwater welding and cutting. It is well equipped with a relatively modern simulator and diving vessel.[38] The school is planned for 1200 students, a level evidently not reached by December 1984.[39]

C-4: VNIIE GAZPROM—Scientific Research Institute of Economics, Industrial Management and Technical-Economic Information for the Gas Industry (A-U)

This is the MGP umbrella research institute for economic and organizational matters. For the offshore industry the institute seems to cooperate with VNIPI Morneftegaz (C-31), for instance on general problems concerning organization, administration and storage of material resources within the all-union production associations (VPO).[40] VNIIE Gazprom has also been studying the use of brigades in offshore organizations.[41]

C-5: VNIIGaz—Natural Gases Research Institute (A-U)

This institute coordinates the more technically orientated research institutions within the gas industry. It specializes in exploration techniques, gas condensation and gas transport systems.[42] There is not much evidence of this institute taking an active part in offshore development.

C-6: VNII ZARUBEZHGEOLOGIYA—Research Institute for Foreign Geology and Offshore Activity (A-U)

The main task of this institute is to inform the Soviet gas industry about Western technology and industry. It has for example gathered information on foreign technology for drilling in shallow waters.[43] It is reasonable to believe that this institute is supplying information to the 'parus' data base at the specialized information centre in Riga (SITs Morneftegaz), which is run by VMNPO Soyuzmorinzhgeologiya (C-32).

D: GOSKOMGIDROMET SSSR—State Committee for Hydrometeorological and Environmental Control (A-U)

Goskomgidromet is taking part in research and development concerning the environmental conditions for Arctic offshore exploration and development (see C-1). Its work in this field is probably coordinated by the Arctic, Antarctic, and Marine Main Administration.[44]

The two oceanography institutes belonging to Goskomgidromet—GOIN, Leningrad (D-1), and GOIN, Moscow (D-2), are probably the central actors in this field. The Leningrad institute conducts general oceanographical research in the Barents and Baltic Seas. The research projects of GOIN in Moscow 'include ocean dynamics, waves, tides, hydrochemistry, environmental problems and assisting industy with hydrometeorological data'.[45]

Another Goskomgidromet institution is the Arctic and Antarctic Scientific Research Institute (AANII) (D-3) in Leningrad. It specializes in studies 'with stress on ice conditions in the Arctic and northern rivers, long-range weather forecasting, and study of the interaction between ice and ship'.[46]

Goskomgidromet is permanently represented in Murmansk by a local hydrometeorology and environmental control administration (D-4).[47] The research institutes are probably represented in Murmansk by more or less permanent expeditions.

E: AN SSSR—The USSR Academy of Sciences

The Academy of Sciences plays a central role in almost every field of research in the Soviet Union. There have however been claims that the Academy is not sufficiently engaged in the petroleum sector, and that coordination of the multitude of branch research institutions, which are subordinated to several ministries, is lacking to a substantial degree. This criticism in Soviet newspapers has particularly applied to the oil industry in connection with the production problems experienced in this branch. However, to our knowledge the AN SSSR seems to be actively participating in research and development in the offshore sector (see C, C-31). There are probably several AN institutions engaged in this work, one of the more important being the Shirshov Oceanology Institute.

E-1: IOAN im. P. P. SHIRSHOVA—The Shirshov Oceanology Institute, Moscow

This is the 'largest Soviet institute for marine science; research involves development of oceanographic methods and equipment. It is also active in use of manned and unmanned submersibles.'[48] As of 1976 this is the principal research agency conducting geological and physical research on petroleum structures of the Arctic shelf, including continuous echo-sounder seismic profile surveys in the northeast Barents Sea, resuming this work in cooperation with Mingazprom in the South of the Barents Sea in 1980–2.[49]

E-2: INSTITUT EKONOMIKI OKEANA DALNEVOSTOCHNOGO NAUCHNOGO TSENTRA AN SSSR—Institute of Ocean Economics, the Far East Scientific Centre of the USSR Academy of Sciences

This institute, situated in the Far East on Sakhkalin island, has conducted studies on the development and running of transport services for the offshore industry.[50]

F: GOSPLAN SSSR—The USSR State Planning Committee

Gosplan is one of the most central and influential organizations in the Soviet economy. Its task is to work out the five-year plans and detailed annual plans that regulate distribution of all material resources and services in the official economy, and all transactions between state economic actors. Gosplan is also engaged in planning concerning foreign trade. Consequently, several departments within the Gosplan central apparatus are concerned with allocating resources to the offshore industry and between different offshore areas. The petroleum and gas department is naturally heavily involved in this.

Gosplan's vice-chairman responsible for this field, A. M. Lalayants, who has been in office for more than twenty years, has shown a clear interest in the question of developing the Barents Sea. In 1984 he visited Norway in order to get acquainted with the Norwegian petroleum industry and what it can offer in offshore related technology and services. Compared with the gas ministry (C), the state committee of science and technology (K) and the ministry of foreign trade (M) Gosplan plays a less prominent part *vis-à-vis* foreign interests seeking access to the possible Barents Sea offshore adventure. That is one of the reasons why it is difficult to say which interests and options Gosplan is favouring: on- or offshore; the Barents Sea or other offshore areas; foreign or domestic technology in Arctic offshore development. It is reasonable to believe, however, that the bureaucratic organization of Gosplan tends to favour traditional incrementalistic solutions, for which results are known in advance, and hence is not a strong supporter of the Barents Sea option.

G: MINGEO SSSR—The USSR ministry of geology

The ministry was responsible for offshore activities until 1978. Presently there seems to be little cooperation between Mingeo and Mingazprom (C), although there is a relationship between subordinate organizations, e.g. VNII Morgeo (G-1) and VMNPO Soyuzmorinzhgeologiya (C-32).

G-1: VNII MORGEO—Marine geology and geophysics research institute, Riga (A-U)

This institute is evidently engaged in conducting or analyzing seismic surveys—or both. It 'explores for mineral oil and gas deposits on the ocean floor and determines the best location for mining or drilling these resources'.[51] In cooperation with the scientific production association NPO Parma in Riga, VNII Morgeo has developed the two data banks which form the backbone of the offshore information centre, SITs Morneftegaz, which belongs to VMNPO Soyuzmorinzhgeologiya (see C-32).[52]

G-2: NPO SEVMORGEO—The Northern Marine Geological-Geophysical Scientific Production Association, Leningrad

We have also found the name Sevmorgeologiya and suppose that this is the same oragnization. It 'has conducted geological and physical research on the Arctic shelf and its oil and gas structures for many years'.[53] In 1984 the institute published the book *Oil and Gas in the World Oceans*.[54]

The Oceanography Research Institute, or VNIIO (A-U), which is also situated in Leningrad, is subordinated to NPO Sevmorgeo. It concentrates on terrestrial geology and geophysics of the Arctic and Antarctic. It makes tectonic maps to forecast mineral resources in polar regions.[55]

H: MINMORFLOT SSSR—The USSR ministry of the maritime fleet

As mentioned in Chapter 2 the ministry of the maritime fleet is providing additional transportation services to the offshore industry in Murmansk. Murmansk Steamship Company (H-1) belongs to the northwest liner department of the ministry. It is reasonable to believe, however, that Minmorflot want to minimize this cooperation, especially during the peak navigation season of the northern sea route going from Murmansk and eastwards. The Murmansk Steamship Company organizes the bulk of the voyages on this route, totalling about 400 ships and 650 voyages in 1985.[56] The company's services are certainly in short supply for the offshore needs, and the Mingazprom is therefore building up its own fleet to become more self-sufficient.[57]

I: MINVUZ SSSR—Ministry of higher and secondary specialized education (U-R)

I-1: MINKhiGP im. GUBKINA—The Gubkin Petrochemical and Gas Industry Institute, Moscow

This is the principal educational institute in the area of oil and gas technology. As already mentioned several crew members of the drillship *Viktor Muravlenko* were reported to have been educated at the Gubkin Institute. The institute also conducts research, but does not seem to be of very much importance in offshore related research. The Minvuz institutions are traditionally less well off than the branch institutes and the institutes belonging to the Academy of Sciences. They are therefore in a less favourable position to carry out expensive research programs, such as those required in the offshore sector.

The professional-technical school in Murmansk (C-37), which belongs to the regional offshore ogranization Arktikmorneftegazrazvedka (C-35) is formally a part of the Mingazprom organization and thus probably able to extract more resources. It is most probable, however, that Minvuz and its Gubkin institute have a say when it comes to the educational programme of the school.

I-2: MISI im. V. V. KUYBYSHEVA—The Kuybyshev construction engineering institute, Moscow

The urban construction section of this Minvuz institute has conducted research on settlement and transportation services in areas with offshore oil and gas production. The emphasis in this study was on the conditions in arctic offshore production regions, with examples from Sakhalin.[58]

J: GOSSTROY SSSR—The USSR State Committee for Construction Affairs

This state committee has coordination responsibilities for the construction sector in the Soviet Union. However, its jurisdiction overlaps with those of several other state and party institutions, and its powers are therefore quite limited. Construction constitutes a serious problem in the Soviet economy because most projects usually take too much time to complete. This leads to a constant retention of enormous amounts of capital in uncompleted construction projects. Another problem is the low quality of many completed constructions. The Soviet economic leadership has therefore on several occasions preferred to use foreign contractors to build important structures on Soviet soil that require quality and a fast completion. This applies both to industrial and other projects such as hotels for foreign tourists.

In the offshore sector Gosstroy participates in planning, research and development. It is most certainly occupied with the construction of an offshore base (J-2) and a new urban centre on the west side of the Kola fjord opposite Murmansk. There are also plans to build a new bridge across the fjord north of the existing one, closer to the two settlements. These are of a traditional design, which hardly require much research. Issues concerning priority, engineering, resource allocation and possible foreign involvement are probably in the forefront in this connection.

Geographically Norwegian, Finnish and Swedish firms are most favourably located for participating in construction of the offshore base. Finnish construction firms have already been engaged in Murmansk; to our knowledge Norwegian and Swedish have not. Foreign participation in construction works is in our opinion dependent upon the urgency of completing the base and on the Soviet plans for a Western presence in the Barents Sea offshore activity generally and on the base itself specifically. This possibly should not be excluded, especially not for the Finns. The Soviet ministry in charge of the construction of the base itself is probably the ministry of construction of petroleum and gas industry enterprises (A-U). This has so far not been confirmed by our sources, however. The construction of the base has been under way at least two years, but we understand that progress has been insignificant.

The other field in which Gosstroy seems to be engaged is in research on and development of ice-resistant concrete production platforms for Arctic regions. There is a variety of institutions taking part in this activity. Gosstroy represents the domestic expertise on onshore concrete constructions. It is an open question to what extent this knowledge can be applied to offshore platforms and which interests Gosstroy has in this field. Soviet publications show, however, that the design work is under way.[59] But it is not known whether the Soviets are conducting this research and development work primarily in order to evaluate offers from foreign companies or to choose their own domestic designs. Promstroyproekt, the State Institute for Industrial Construction Design (J-1), which we believe belongs to Gosstroy, is taking part in this research and development work.

K: GKNT SSSR—The USSR State Committee for Science and Technology

As the name indicates, this state committee has a very broad field of responsibility. Its primary task is to enhance the technological level in all spheres of Soviet production. To serve this end it has quite widespread contacts abroad, being represented in all major Soviet embassies and trade representations. The organization's task is to map the needs for new production methods and machinery, and to try to meet these needs by keeping track of what is happening in the different fields of Soviet and foreign research and technology. If the Soviet Union cannot supply the item in question, the GKNT will evaluate foreign alternatives. GKNT is usually not actively engaged in traditional purchases of standard products from abroad. But it is supposed to play an active role when technology transfers, production cooperation or compensation deals are in question.

Under Gorbachëv GKNT has been the object of very strong criticism. In a speech in September 1986 Gorbachëv himself accused GKNT of not being willing to change in accordance with the new party line and plainly spelled out that this organization 'does not as yet fulfil its role'.[60] GKNT has been participating in contacts with Western offshore companies from an early stage of the discussions on offshore cooperation and technology. It is worth noticing that GKNT is the formal counterpart to the Norwegian offshore consortium BOCONOR in the joint working group that has been established (O).

L: MID SSSR—USSR ministry of foreign affairs

Up to 1986 MID apparently has not played an active role in connection with the Barents Sea contacts with Western offshore companies. It is clear, however, that its former Scandinavian department, and now the 2nd European department[61], have been kept well informed, and are considering negotiations and different cooperation concepts in an overall foreign policy perspective. By May 1986 MID officials would not totally exclude a concept for developing the Barents Sea oil based on the North Sea model, i.e a concession system also open to foreign companies, but considered it improbable[62]. When and if more specified cooperation concepts materialize, one must expect that not only MID, but also the ministries of defence and of the fish industry will start taking a more active part in the internal Soviet decision-making process. It is obvious that large-scale decisions, as depicted in Chapter 4, will involve MID to a substantial extent.

M: MINVNESHTORG (MVT) SSSR—USSR ministry of foreign trade

There are three administrations in MVT concerned with Barents Sea offshore development: the trade sector main administration for import of machinery and equipment from capitalist countries, the functional main administration for compensation projects with capitalist countries and the commercial and political administration for trade with Western countries.

Sushkov's position as Soviet chairman in the joint Norwegian-Soviet trade Commission has been taken over by Vladimir Burmistrov, who was probably appointed deputy minister in early 1986. It is uncertain whether Burmistrov now has the overall responsibility for the Barents Sea within the ministry because other deputy ministers handle trade with Finland, Sweden and other European countries that could be interested in Barents Sea offshore activities. The day-to-day handling and coordination of the Barents Sea issue within the ministry seems to be conducted by the administration for trade with Western countries, headed by Yuri P. Ledentsov.[63]

When it comes to the technicalities and organization of a big compensation deal for the Barents Sea the main administration for compensation projects with capitalist countries will play a central role. To our knowledge this administration, with its chief Andrey Ponomarev, has already been engaged in preliminary work in this field. The main administration for import of machinery and equipment from capitalist countries is probably responsible for more limited puchases of rigs, vessels and other offshore equipment. The formal Soviet trading partner in these purchases is, however, the trade organization VO Sudoimport (M-1). The firm Sudoshelf, which is a part of this organization, possesses the commercial expertise to conclude contracts in the offshore field of business (see Appendix II). The technological expertise is found, however, within the Mingazprom organization (C, C-3; e.g. C-21, C-31, C-32, C-35), and to a lesser extent within GKNT (K).

N: Soviet–Finnish Arctic technology committee

Soviet–Finnish cooperation in shipbuilding for use in Arctic regions has a long tradition. Soviet and Finnish ice-breaker technology is world leading. Partners in this trade have been the well known Finnish shipbuilders Rauma Reppola, Wartsila, Hollming and Valmet and the Soviet trade organization V/O Sudoimport (M-1). This cooperation has now developed into the new field of Arctic offshore exploration equipment; it comprises drilling ships, semisubmersible and jack-up drilling rigs and other vessels. Four of the drilling units already operate in the Barents Sea (see Chapter 2).

As early as 1977 the temporary committee 'Arctic Project' was formed within the framework of the permanent intergovernmental economic committee. It has been working on a permanent basis since

1981 under the name 'Arctic Technology Committee'. On the Finnish side there are representatives from companies in shipbuilding, construction and metal industries, from governmental research and trade organizations. The Soviet side is represented by the gas ministry, the state committee for science and technology and some central research and design institutions.

The objectives of the committee are to find solutions to several technological problems connected with Arctic development. The research is divided into three main areas:

1. To supply information on critical environmental factors that have to be taken into account when constructing equipment and facilities for use both on- and offshore in Arctic areas. Special attention is given to ice conditions and the seabed.
2. To design artificial islands, steel and concrete platforms for oil and gas drilling and/or production. Marine constructions in ice covered waters are of special interest to the committee.
3. Technical support systems for production, handling and transportation—both on- and offshore; concentrated coast bases are of special interest.

Thirty to forty years of man labour were carried out on the Finnish side in 1984. In the committee's new programme for 1986–90 sub-sea technology will be given special attention. In addition to the committee, an 'offshore work group' was established in 1983 to promote and coordinate practical results of research.[64]

O: GKNT-BOCONOR joint working group

In April 1983 a contract was signed between Sudoimport (M-1) and Norwegian Petroleum Consultants A/S for the delivery of 'a masterplan for a Barents Sea project'. The contract was signed in the presence of deputy foreign trade minister Sushkov (M) and the Norwegian trade minister Arne Skauge and attracted much publicity.

During the summer of the same year a consortium was established, consisting of seven leading Norwegian offshore companies, including Norwegian Petroleum Consultants. Barents Sea Offshore Consortium of Norway (BOCONOR) could offer the whole range of offshore equipment and services, and was primarily presented as a turn-key contractor for a possible Barents Sea project. Since the establishment of the consortium, it has been the main theme of official Norwegian–Soviet trade discussions. BOCONOR has participated in trade fairs in the Soviet Union and received several Soviet delegations in Norway. As a result an agreement with GKNT (K) was concluded in late 1984 or 1985 and a joint working group was established on the basis of this agreement. It foresees exchange of scientific and other information in order to find the best possible solutions for offshore activities under Arctic conditions.

P: MINSUDPROM (MSP) SSSR—USSR ministry of the shipbuilding industry

This ministry is engaged in both civilian and military shipbuilding[65]. It is probably responsible for the domestic offshore rigbuilding, which is concentrated in Astrakhan and Baku by the Caspian Sea and in Vyborg by the Baltic. The Vyborg yard (P-1) is producing ice-resistant semi-subs of the shelf-model for Arctic regions. In 1985 the first of these were put into operation in the Barents Sea. The Vyborg yard will in cooperation with the Finnish firm Rauma-Reppola also construct two large Arctic jack-ups by 1987–8 (see Chapter 2, Appendix 2, for more information).

Q: MINKHIMMASH SSSR—USSR ministry of chemistry and petroleum machine-building

The ministry 'produces drilling equipment of all types, rock-breaking and core-sampling equipment, drilling and casing pipe, drill collars, and down-hole equipment. It also supplies petroleum production equipment and automation machinery. . .'[66] To what extent the equipment produced by this ministry is used in the offshore sector is not known to us, but its participation in offshore-related research, development and production is well established.[67]

Notes

1. *Directory of Soviet Officials: National Organizations* (DSONO), Washington D.C., Directorate of Intelligence, November 1984, p. 13 (I) and August 1983, p. 13 (II).
2. Boris Shcherbina was for several years party secretary in Siberia, first in the Irkutsk oblast party committee (1951–61), and then first secretary in Tyumen oblast (until 1973). He was then appointed minister for the construction of petroleum and gas industry enterprises (Minneftegazstroy). As a party official he actively supported Siberian interests, lobbying for the development of the oil and gas fields in the Tyumen oblast. As minister he showed unusual organizational talents when responsible for the construction of the gas pipelines running from Western Siberia to the European part of the USSR and Western Europe. Shcherbina was appointed head of the *ad-hoc* commission to take care of the accident at the Chernobyl nuclear plant in April 1986.
3. In November 1985 GOSAGROPROM (state committee for the agro-industrial complex) was established by merging five ministries and one state committee.
4. In the machine-building sector this new coordination bureau is headed by SOVMIN deputy chairman I.S.Silayev (appointed November 1985). (RL 145/86, *Radio Liberty Research*, 1 April 1986, p. 19.)
5. For more information on this bureau of the USSR SOVMIN see Chapter 4, note 13.
6. See also Chapter 2 and note 37 to that Chapter.

7. *Aftenposten*, 2 July 1985.
8. *DSONO* (I) p. 106 and (II) p. 137.
9. Ibid., p. 107(I).
10. *Soviet Shipping*, 3, 1983, pp. 17–19. In this article Nemchinov describes the offshore fleet.
11. *DSONO*, op. cit., p. 106(I).
12. Derezhov has been a member of the joint governmental Norwegian-Soviet commission for economic cooperation, where the Norwegian offshore consortium BOCONOR's offers have been discussed. He also presented a paper on offshore cooperation between the USSR and the Nordic countries at a symposium in Moscow in September 1985. ('Sotrudnichestvo v oblasti energetiki; po materialam mezhdunarodnogo simpoziuma', *Ekonomicheskaya gazeta*, 40, 1985, p. 22.)
13. R. S. Vorobeva: 'Stroitelstvo neftegazopromyslovykh gidrotekhnicheskikh sooruzheniy v usloviyakh shelfa', *Gazovaya promyshlennost*, 4, 1984, p. 43; R. S. Vorobeva: 'Gidrometeorologicheskie parametry dlya sozdaniya tekhnicheskikh sredstv osvoeniya shelfa', *Gazovaya promyshlennost*, 8, 1985, p. 37.
14. In a speech at a MGP meeting, 17 October 1985, the gas minister Chernomyrdin urged the geological administration to help in speeding up exploratory drilling. 'Dostoyno vstretit partiyny forum', *Gazovaya Promyshlennost*, 12, 1985, pp. 1–3.
15. Meeting with regional managers, December 1984.
16. See Chapter 2, for an evaluation of Soviet ability to process seismic data from the Barents Sea continental shelf.
17. NII Morgeofizika was one of the MGP organizations scoring highest on the inventiveness indicator for 1983. (*Gazovaya promyshlennost*, 6, 1984, p. 27.)
18. This happened to the expedition which preceded the all-union design institute Spetsmorneftegazproekt (C-34) ('40 let proektirovanya', *Gazovaya promyshlennost*, 3, 1985, p. 13).
19. *Gazovaya Promyshlennost*, 12, 1985, p. 3 (see note 14).
20. Numerous articles on offshore matters in *Gazovaya Promyshlennost* during the last two and a half years.
21. 'Funktsionalnaya skhema sluzhby glavnogo svarshchika Mingazproma', *Gaz.prom.*, 4, 1985, p. 15.
22. Cooperation with Promstroyproekt (J-1) on price calculations for different kinds of arctic offshore installations: 'Sozdanie neftegazopromyslovykh gidrisooruzheniy dlya shelfa severnykh akvatoriy', *Gaz.prom.*, 5, 1984, p. 40. Prognostication of the economic effects of offshore development: 'Prognozirovanie narodnokhozyastvennogo effekta osvoeniya mestorozhdeniy shelfa,' *Gaz.prom.* 12, 1985, pp. 30–32. Concrete Arctic offshore installations: study trying to assess the relative efficiency of different solutions based on a set of given criteria. Price calculations. ('Zhelezobetonnye neftegazopromyslovye sooruzheniya dlya arkticheskogo shelfa,' *Gaz.prom.*, 12, 1985, pp. 32–3.)
23. 'V bumazhnykh zavodakh,' *Sotsialisticheskaya industriya*, 17 July 1985. According to this article a superior Soviet design for an offshore pipelaying vessel was scrapped in favour of Western imports.

24. The geological and geophysical institutes are most likely not represented in this council.
25. 'V nauchnom tsentre morskoy podotrasli', *Gaz. prom.*, 6, 1984, pp. 45, 8, 1985, p. 37 (see note 13).
26. 'Kolonka revizora. Vskryty fakty pripisok', *Gaz.prom.*, 11, 1984, p. 21.
27. See Chapter 2.
28. 'Vedomstvennye stroitelnye normy. Inzhenernye izyskaniya na kontinentalnom shelfe', *Gaz.prom.*, 8, 1985, p. 49.
29. 'Avtomatizirovannye sistemy informatsionnogo obespecheniya dlya morskikh neftegazovykh promyslov', *Gaz.prom.*, 1, 1986, p. 13. The article gives more detailed information on how the information is organized and distributed. It also gives the names of the EDP hardware used.
30. *Gaz.prom.*, 9, 1984, p. 11.
31. *Gaz.prom.*, 10, 1984, p. 38.
32. '40 let proektirovaniya', *Gaz.prom.*, 3, 1985, pp. 13–14 (see note 18).
33. 'Sovershenstvovanie ledostoykikh konstruktsiy morskikh neftegazodobyvayushchikh platform', *Gaz.prom.*, 8, 1985, pp. 42–5.
34. 'Kakim byt upravleniyu podryadnoy stroitelnoy deyatelnostyu' *Gaz.prom.*, 8, 1984, p. 19.
35. 'Rasplata za pripiski', *Gaz.prom.*, 6, 1984, p. 41.
36. Article where Arktikmorneftegaz is mentioned in another context: *Sotsialisticheskaya industiya*, 4 February 1986.
37. *Gaz.prom.*, 3, 1984, p. 7.
38. *Gaz.prom.*, 4, 1985, p. 15.
39. According to regional managers in Murmansk, December 1984.
40. *Gaz.prom.*, 9, 1984, pp. 36–8.
41. *Gaz.prom.*, 9, 1984.
42. *Directory of Soviet Officials: Science and Education* (DSOSE), Directorate of intelligence, Washington D.C., August 1984, p. 292.
43. *Gaz.prom.*, 10, 1984, pp. 44–7.
44. *DSONO*, November 1984, p. 195.
45. *DSOSE*, September 1985, p. 315.
46. Ibid., p. 311.
47. *DSONO*, November 1984, p. 196.
48. *DSOSE*, August 1984, p. 135.
49. Private source, November 1985 (reference to *Pravda* 5 October 1983).
50. *Gaz.prom.*, 10, 1984, p. 39.
51. *DSOSE*, September 1985, p. 294; and *Gaz.prom*, 10, 1985, p. 38.
52. *Gaz.prom.*, 1, 1986, p. 13.
53. Private source, November 1985.
54. *Neftegazonosnost mirovogo okeana*, Leningrad, 1984.
55. *DSOSE*, September 1985, p. 295.
56. *Polar Record*, 23(143), 1986, p. 186.
57. N.Nemchinov, 'Fleet for Offshore Oil and Gas Business'. *Soviet Shipping*, 3, 1983.
58. *Gaz.prom.*, 2, 1985, pp. 33–4.
59. *Gaz.prom.*, 5, 1984, pp. 40–1.
60. *Sotsialisticheskaya Industriya*, 20 September 1986, p. 2.
61. *Aftenposten*, 13 August 1986, p. 9.
62. Informal conversation with MID officials, May 1986.

63. *DSONO*, June 1986, p. 82.
64. Pauli Jumppanen, paper to symposium in Moscow, September 1985.
65. *DSONO*, June 1986, p. 147; and W. J. Kelly, H. L. Shaffer & J. K. Thompson, *Energy Research and Development in the USSR*, Durham, Duke University Press, 1986, p. 128.
66. *Energy Decision Making in the Soviet Union*, Directorate of Intelligence, Washington D.C., 1982; Organizational chart.
67. Kelly *et al.*, op. cit, pp. 127–8.

Appendix 2
Soviet offshore-related purchases from abroad

In order to demonstrate to what extent the Soviet Union is buying foreign offshore equipment, we have compiled a list of such purchases. The information has been gathered from scattered sources and the list is certainly not complete. The predominance of the Finnish shipbuilders in this list does, however, seem to reflect the reality of Soviet imports of maritime equipment, especially when it comes to Arctic areas. In addition to Finland, Poland is also an important supplier of different kinds of vessels. The Soviet Union and Poland have concluded an agreement on production of offshore related shipbuilding, the so-called 'shelf' programme[1] which includes a substantial number of tug and supply vessels. Besides Finland and Poland we also find Romania, West Germany, Norway, The Netherlands, Brazil, Japan and Singapore among the suppliers of offshore rigs and vessels.

In March 1983, deputy minister of the gas industry, N. Nemchinov, mentioned that the ministry had acquired forty ships during the preceeding two years, and that more than 300 additional vessels are planned for the future. He did not say, however, how much of this will be of foreign origin.[2] According to the trade agreement between the USSR and Finland for the current five-year plan period (1986–90), exports of Finnish shipbuilders to the Soviet Union will equal 2.6–3 billion roubles ($3.6–4.2 billion, or FIM 19–22 billion[3]). This is 20 per cent less than in the last five-year plan period.[4] As our list shows, a substantial share of this is already contracted. But as a result of the dramatic fall in oil export revenues one may expect cuts in the purchases from Finland.[5]

Notes

1. PAP in English, 1700 GMT, 11 October 1984, *BBCMS BBCEE*—BBC Monitoring Service, Eastern Europe (data base abstract).
2. *Soviet Shipping*, 3, 1983, p. 17.
3. According to the USSR state bank's official exchange rates from 2 June 1986: 100 USD = 71.70 SUR; 100 FIM = 13.71 SUR.
4. *Affarsvarlden*, 9 April 1986, pp. 95–7 (data base abstract).

5. *Financial Times*, 26 April 1986, reported that the USSR is likely to cut plans for imports of petrochemical equipment substantially. Energy equipment is, however, expected to be among the high-priority imports.

1. Mobile drilling rigs

Contract	Number	Description	Companies/ Country	Delivery/ Value	Source
?	1	*Khazar* Jack-up drilling rig in the Caspian Sea. Prototype for the Soviet *Kaspii* series.	Netherlands	1974 ?	*OGJ*,* 12 October 1981, p. 168
?	1	*Okha* Jack-up drilling rig, Sakhalin; operator: Sodeco. (83–5)	Mitsu Japan	1980 ?	*Petroleum News*, December 1985
?	1	*Kaspmorneft* (partially assembled in Astrakhan) semi-sub rig; Caspian sea; prototype for the Soviet series of 'shelf' semi-subs	Rauma-Reppola Finland	1981 ?	*PE*† December 1985
1979	3	Drillships *Valentin Shashin* *Viktor Muravlenko* *Mikhail Mirchink* Barents Sea (2) and Sakhalin/Vietnam (1) Norwegian dynamic positioning equipment diving station	Rauma-Reppola Finland	1981–2 FIM 900 million	*OGJ*, 26 April 1982, p. 93; *PE*, April 1984, p. 146

?	1	*Ekhabi* Jack-up; Vietnam and off Sakhalin. Operator: Viet-Sovpetro	FE Levingston Singapore	1984 ?	*Petroleum News*, December 1985, p. 146
?	1–2	Jack-up drilling rig, probably for use in the Black Sea	? Constanta, Romania	Under construction, 1984 ?	*PE*, April 1984, p. 146
January 1983	2	Jack-up rigs: *Kol'skaya* and *Sakhalinskaya* 15 000 tonnes, 100/6,500 m.	Rauma-Reppola Sudoimport Finland	1985 US$180 million	*OGJ*, 4 February 1985; *PE*, July 1985
1984	4 (6)	Jack-up rigs, rig components and various equipment. MK-Holland design (Option for 6 more)	Petrobras Brazil	1988–9 US$750 million	*OGJ*, 25 March 1985, p. 68
March 1985	2	Equipment, parts for world's biggest jack-ups, 140 metres. (Also cooperation agreement signed.)	Rauma-Reppola Sudoimport Finland	1986–7 FIM 300 million	*Hbl*,§ 23 March 1985; *OGJ*, 15 April 1984, p. 40

* *Oil & Gas Journal*
† *Petroleum Economist*
§ *Hufvudstadsbladet*, Helsinki.

2. Rescue, fire-fighting and spill containment vessels

Contract	Number	Description	Companies/ Country	Delivery/ Value	Source
?	1 (probably more)	Ice-breaking tow and fire-fighting vessel; *Deymos*. Length: 71.6 m, width: 18 m; Complex diving station.	Wartsila/ Sudoimport Finland	Spring 1984 ?	*Gaz. prom.*,* 4, 1984, p. 45
?	1	Multi-purpose vessel for spill containment, discharge processing, fire fighting and dredging. Spill containment capacity: 800 m^3/h. Capacity of each sea-bed pump: 13 000 m^3/h (10,500 dwt *Gogland* may be precedent).	Wartsila/ Mingazprom Finland	under construction, January 1985 ?	*Gaz. prom.*, 1, 1985, p. 60

April? 1984	2	Arctic class rescue tug 'SB-922'	Rauma-Reppola/ Sudoimport	1985–6	*Kauppalehti*,† 13 April 1984, p. 1; 15 May 1985, p. 13 †
			Uusikaupunki, Finland	?	
May 1986	4	Oil-spill recovery vessel	FELS/ Sudoimport	1987	*OGJ*, 2 June 1986, p. 35
			Singapore	US$24 million	

* *Gazovaya Promyshlennost*, Moscow.
† Data base abstract.

3. Crane and heavy lift vessels

Contract	Number	Description	Companies/ Country	Delivery/ Value	Source
September 1981	1	26,800 dwt heavy-lift vessel for transporting and locating drilling rigs up to 18 000 tonnes in the Caspian Sea.	Blohm & Voss BRD	? DM 75 million	*Handelsblatt*, 23 September 1981, p. 15.
?	4	Gigantic 'titan' crane catamarans; offshore construction; lifting capacity 600 tonnes; diving station; to perform several kinds of offshore operations. Production cooperation on the two assigned for the Caspian Sea.	Wartsila/ Sudoimport Finland	1984–5 (84: Titan-1; Autumn 1985: Titan-2) ?	*Gaz. prom.*, 1, 8, 1986; 6, 1985.

?	1	Crane vessel 'Atlet-4', lifting capacity: 40 tonnes; to deliver equipment for off-shore drilling rigs. Delivered to 'Kaliningradmorneftegaz-prom', the Baltic Sea.	Poland Gdansk, Poland	January 1986 ?	*BBCMS, BBCSU*, 11 January 1986.*
?	1	Crane vessel, 50,000 tdw; lifting capacity 1600 tonnes to 60 metre height. (World's largest.) Diving complex for sub-sea works down to 50 m.	Wartsila/ Sudoimport Turku, Finland	August 1985 ?	*Gaz prom.*, 1, 1986, p. 50; *Kauppalehti*, 1 August 1985, p. 4†
July 1985	1	Heavy-lift vessel; 34,000 dwt; for transportation and docking of mobile jack-up and semi-submersible drilling rigs up to 20 000 tonnes.	Wartsila/ Sudoimport Turku, Finland	Early 1987 FIM 250 million	*Gaz. prom.*, 1, 1986, p. 47; *Kauppalehti*, 25 July 1985, p. 4.†

* BBC Monitoring Service, Soviet Union; Data base abstract.
† Data base abstract.

4. Accommodation and crew tender vessels

Contract	Number	Description	Companies/ Country	Delivery/ Value	Source
August 1981	4	Accommodation vessel for 394 people	Valmet	1983–4	*Lloyds List*, August, 35, 1981.*
			Finland	?	
?	1	*Kalevala* vessel serving as living quarters for the Murmansk based offshore workers.		1984	Private source, November 1985
			Finland	?	
April 1984	3	Houseboat	Valmet	1985–6	*Kauppalehti*, 16 April 1984,* p. 6.
			Turku, Finland	FIM 250 million	
May 1985	5	Crew tender, to take crews to drilling platforms	Centromar/ Sudoimport	1988	*BBCMS, BBCEE*.†
			Gdynia, Poland	?	

* Data base abstract.
† BBC Monitoring Service, Eastern Europe; data base abstract

5. Supply and other service vessels

Contract	Number	Description	Companies/ Country	Delivery/ Value	Source
August 1981	4	Offshore support vessel	Valmet Finland	1983–4 ?	*Lloyds List*, 25 August 1981.*
?	7	Supply vessels, UT 700, 2 in Barents	Ullstein/ Sudoimport Finland	1976–84	'Neftegaz-1984'†
October 1984	25	Offshore support vessels, based on Ullstein's 706 design (Tugboats?)	Warski yard Poland	1986–90 SUR 310 million	*Lloyds List*, 25 October 1984, p. 1.* *BBCMS-EE*§
?	20	Supply vessels	Gdansk Poland	under construction 1984 ?	*PE*, April 1984, p. 146.
January 1985	2	Cable vesel, 800 tonnes. Prolonged model of *Emba*, 86 by 12.5 metres	Wartsila/ Sudoimport Finland	1986 FIM 200 million	*Kauppalehti*, 10 January 1985,* p. 16.

* Data base abstract.
† Brochure from The Export Council of Norway at the offshore exhibition 'Neftegaz-84', Moscow, 27 November–4 December 1984.
§ BBC Monitoring Service, Eastern Europe; data base abstract

6. Research vessels

Contract	Number	Description	Companies/ Country	Delivery/ Value	Source
August 1981	5	Special ocean research vessel	Valmet	1983–4	*Lloyds List*, 25 August 1981.*
			Finland	?	
April? 1984	2	Vessel for charting under-water topography	Rauma-Reppola	?	*Kauppalehti*, 13 April 1984,* p. 1.
			Finland	?	
October 1984	4	Geophysical vessels	Wisla yard	1986–9	*Lloyds List* 15 October 1984*
			Poland	SUR 21 million	
February 1985	2	Geophysical research vessels for shallow waters (3–8 m), oil and gas finding in Kara, Barents and Caspian Seas; ice class L1; 49.8 by 10.5 metres; 2 metre depth.	Rauma-Reppola Sudoimport	1987–8	*Hbl.*† February 1985
			Finland	FIM 100 million	

?	3	Geological research vessels, 2600 tonnes.	Hollming Finland	1985–6 ?	*Veritas*, July–August 1984.§
1984	2	Geological research vessels, 5300 tonnes; operate on depths down to 300 metres; drill 200 metre holes in the sea-bed. Dynamic positioning system.	Hollming/ Mingazprom. Finland	1986–7 ?	*Veritas*, July–August 1984.
April?	2	Research vessel (the contract not finally confirmed in our sources).	Hollming Finland	End 1980s FIM 400 million	*Kauppalehti*, 10 April 1985, p. 5.*
January 1985	4	Hydrometeorological research vessel	Valmet/ Sudoimport Finland	1986–7 FIM 150 million	*Hbl.*† January 1985.
May 1986	3	Research vessel (similar to four vessels built previously by Valmet in Turku).	Valmet Finland	? FIM 100 million	*Kauppalehti*, 21 May 1986, p. 1.*

* Data base abstract.
† *Hufvustadsbladet*, Helsinki.
§ Periodical published by the Norwegian company Det Norske Veritas.

Index

(Index does not include details on equipment and companies from appendix2)

Academy of Sciences of the USSR 109, 111, 115
  branch in Novosibirsk 22, 23
  Far East Scientific Centre of the Academy of Sciences 115
active sonar 71, 83
Aganbegyan, Abel 23
agriculture 7
ALFA submarines 67
Amerada 47
American forward-based naval strategy 63, 82
American Pacific 79
American sanctions 25
American technology 25, 96
AN SSSR *see* Academy of Sciences
Anadyr 25
Andøya 47
Antarctic 116
anti-submarine warfare, *see* ASW
April, 28th, field of 23
Arkhangelsk district 26
Arctic and Atarctic Scientific Institute (AANII) 114
Arctic Ocean 58, 63, 64, 69, 77, 79, 81, 82, 84
Arctic Technology Committee, *see* Finland
Arktikmorneftegazflot 44, 113
Arktikmorneftegazrazvedka 44, 50, 108, 110, 113
  professional-technical school 113
Arktikugol 56
Astrakhan 37, 128
ASW 64, 67
  aircraft carrier Kiev 67
  equipment 70
  forces 68, 70, 76, 82
  operations 76, 79
  preparedness 68
  tasks 67
  tests 68
Atlantic Fleet 79
attack submarines 68, 69, 70, 83, 96

background noise 96
Baku 1, 112
balance of terror 63, 64, 68
Baltic Sea 26, 37, 88
Barents Base Kirkenes 56
Barents Sea Offshore Consortium of Norway, *see* BOCONOR
Bear Island 59, 65, 72, 80, 81
  Bear Island South 46
Bergsager, Egil 54
Bering Sea 79
Bering Strait 79
Black Sea 26, 129
Boconor 44, 119, 121
Brazil 126
BRD 104, 132
British gas market 101
British Petroleum 47
Burmistrov, Vladimir 120
Byely Island 26

capital investments 93, 95
Caspian Sea 1, 23–4, 27, 38, 110
catamarans 38, 133
Central Asia 23, 27
Central Committee 106
Central Committee Secretariat 107
Chaivo field 24
Chebrikov, V. 106
Chernobyl 15–16, 122
Chernomyrdin, V. S. 51, 56, 108, 123
China 63
Chuckchi Sea 79
CIA 2, 4, 104
clearing (trade) 89

CMEA 88
coal 1
Cocom 100
Communist Party of the Soviet Union, *see* KPSS
compensation deals/projects 91, 95, 96, 97, 99, 120
computer equipment 51
concrete platforms 92
Conoco 47
conservation 6, 14, 19
Constanta 129
construction, *see* Gosstroy
costs 13, 18, 22, 93, 94
Council of Ministers of the USSR 107
Crimea 26
Critchley, Harriet 71, 76
crude sharing 100

Delta submarines 69, 72, 76, 80, 86
demarcation line 33, 35, 36, 77
Derezhov, S. R. 108
Dinkov, V. A. 31
disputed area 33, 35, 36
Dolgikh, V. I. 106
drilling equipment 53
drilling record 50
drilling rigs 50, 53, 56
drillships 35, 50
dynamic positioning equipment 36, 39, 50, 128

East Greenland Sea 69
Eastern Europe 9, 11–12, 88
Eastern Siberia 22–3, 27
Ekonomiki Okeana Dalnevostochnogo Nauchnogo Tsentra AN SSSR, Institut 115
electricity 14
Elf 47
embargo 100–2
energy conservation, *see* conservation
energy consumption 12, 13
energy policy coordination bureau of the USSR Council of Ministers 107
equipment manufacturers 101
Esso 47
European North 26, 27
exploration 12, 20, 21, 27
exports (Soviet)
  arms 5, 6, 10
  gas 10, 91
  gold 6, 10
  oil 3–6
extensive growth 13
external transit 64, 68, 80, 82, 84

Far East 24, 27
Federal Republic of Germany 104, 132
Fina 47
Finland 88, 89–90, 120, 126
  Arctic Technology Committee, Soviet–Finnish 103, 120, 121
Finnmark West 46
fish industry 51, 119
fixed platform 36
floating platforms 23, 50
foreign equipment 53
foreign investors 95
foreign partners 88
foreign policy 96
foreign trade 4–12,
Fram Strait 81
France 104
Franz Josef's Land 58, 59, 81, 85
Frøysnes, Torbjørn 78

gas 13–14
gas deals 91, 98
gas finds 47
gas industry 19, 98
gas ministry 39, 44, 45, 50, 51, 52, 54, 56, 91, 96, 98, 103, 108
  geological administration of the 109
  scientific–technological council 109
  offshore exploration, development and production main organization 44, 52 110
gas negotiations 101
gas pipeline deals 95
gas reserves 47
Gazprom, VNIIE 113
Gdansk yard 38
geological exploration 32, 92, 104, 116
geophysical surveys 104, 116
Giprmorneftegaz, Gos NIPI 110, 112
GIUK gap 68, 69, 77
GKNT–BOCONOR joint working group 121
GKNT SSSR 44, 103, 111, 119
Glavmorneftegaz 44, 52, 110
GOIN 114
  Leningrad 114
  Moscow 114
gold production 10
Golitsyn field 26
Goose bank 35, 36
Gorbachev, M. S. 18, 29, 99, 106, 119

Goryachkin, V. I. 55
Gosagroprom 122
Goskomgidromet SSSR 109, 111, 114
Arctic, Antarctic and Marine Main Administation 114
Gosplan SSSR 16, 45, 115
Gosstroy SSSR 109, 111, 118
Gostekhnadzor SSSR 112
Greenland 81
Gremikha 63
grey zone agreement 66
Gromyko, A. A. 106
Gubkin Petrochemical and Gas Industry Institute 39, 117

Hammerfest 47, 80
hard currency needs 6
hard currency strategy 92–103
Harstad 47
Heavy Industry and Power Engineering department of the Central Committe of KPSS 106
heavy-lift vessel 38, 132–3
Hollming 38, 120
Hopen 59, 81
hydrochemistry 114
hydrometeorology, *see* Goskomgidromet

ice floes 58
ice ridges 58, 59
ice-covered area 58
ice-free area 57–8, 66
ice-bergs 58, 59, 68, 72, 74, 81
ice-resistant concrete production platforms 118
imports (Soviet) 7
food 7, 8
grain 7, 9
machinery and technology 9
Institute of Industrial Economics, Bergen 104
intensive growth 13
interest rate 93
internal transit 63, 64, 68, 80, 84
investments 16–20, 27, 95
IOAN im. P. P. SHIRSHOVA 115
Italy 104

jack-up rigs 35, 37–8, 56, 128–9
Jakutia 23
Japan 24, 25, 101, 126
joint governmental Norwegian–Soviet commission for economic cooperation 123
joint ventures 99

Kaliningrad 26
Kamchatka 25
Kara Sea 26, 32
Kaverochkin field 24
Kazakh republic 108
Kazakhstan 23, 27
Kirkenes 47
Kirov class 67
Kola base 63, 68, 77, 79
Kola Fjord 39, 61, 118
Kolguev Island 26, 35, 37, 44
Kolskaya 35, 37
Komi autonomous republic 26
Kongsberg Våpenfabrikk 36
KPSS 106
Kresta-II 67
Krivak class 67
Kuriles 25
Kuybyshev construction engineering institute, The 117

Lalayants, A. M. 115
Ledentsov, Yu. P. 120
LNG 24

Maksutov, P. A. 111
Mangyshlakneft 108
manoeuvering space 96
Marine geology and geophysics research institute, *see* Morgeo
Marlene Østervold 77, 78, 99
Mikhail Mirchink 36, 128
Middle East 4, 5
middle triassic deposits 35
MID 119
Mingazprom SSSR (MGP) *see* gas ministry
Mingeo 116
ministry of chemistry and petroleum machine building 122
ministry of defence 119
ministry of foreign affairs 119
ministry of foreign trade 91, 96, 98, 120
ministry of geology 116
ministry of the oil industry 98
ministry of the maritime fleet 44, 117
ministry of the shipbuilding industry 122
ministry of construction of petroleum and gas industry enterprises 118, 122
Minkhimmash 122
MINKhiGP im. GUBKINA, *see* Gubkin institute

Minmorflot 44, 117
Minneftegazstroy 118, 122
Minrybkhoz 112
Minsudprom (MSP) 122
Minvneshtorg (MVT) 91, 96, 98, 120
Minvodkhoz 112
Minvuz 109, 111, 117
Minzdrav 112
MISI im. V. V. Kuybysheva, *see* Kuybyshev
Mobil 47
mobile rigs 23, 36–8, 128–9
Morgeo, VNII 112, 116
Morgeofizika, NII 44, 109
Morneftegaz, VNIPI 56, 110
NTTs, The Scientific-Technical Centre
Murmansk 39, 44, 51, 53, 55, 61, 108, 109, 113, 117, 118
Murmansk High 32, 33, 35, 36, 65
Murmansk Steamship Company 117
Murmanskaya, *see* Murmansk high
MVT, *see* ministry of foreign trade

natural gas, *see* gas
naval strategy 97
Netherlands 126
Nemchinov, N. M. 108, 126
Nenets autonomous district 26
NGL 4
Nordic companies 101
Norsk Hydro 47
North Atlantic 79, 84
North Cape 85
North Cape basin 47
North Sea 70
Northeastland 59, 72, 73, 85
Northern Marine Geological-Geophysical Scientific Production Association, *see* Sevmorgeo
northern sea route, the 117
Norway's Geographical Survey 85
Norwegian Barents Sea 46
Norwegian Petroleum Consultants A/S 121
Norwegian Petroleum Directorate 47
Norwegian Polar Institute 85, 86
Norwegian Sea 81, 82
Norwegian shelf 45–9, 67
Novaya Zemlya 32, 63, 72, 81, 85
Novosibirsk 22, 23
NPD 47
nuclear energy 14, 15

oceanography 114
Oceanography Research Institute, *see* GOIN
Oceanology, *see* IOAN
Odoptovskoye 24
offshore base 47, 118
oil consumption 6
oil exports 3–6, 12
  Eastern Europe 5
oil ministry 98
oil price
  domestic 14
  world market 12, 103
oil production 1, 4, 6, 19, 21
  costs 13
oil reserves 2
Ostisty, B. K. 109

pack ice 68, 72
Pacesetter design 37
Parma, NPO 112, 116
party leadership 97–8
passive sonar 67, 69, 70, 71
Pechenga Fjord 61
Pechora river 35, 36
Pentagon 69
permanenet geological–geophysical expedition, Murmansk 109
Peschanoozerskoye oil and gas find 26
pipeline deals 91
pipeline dispute 25, 96, 100, 101, 102
Poland 112, 126
Politburo 106
Polyarny 61
Ponomarev, Andrey 120
power stations 14
Pricaspian Depression 27
production costs 22
professional–technical school, belonging to Arktikmorneftegazrazvedka, Murmansk 113
Promstroyproekt 119
prospecting profiles 32
Ptitsyn, V. N. 44, 55

Rauma Reppola 35, 36, 37, 56, 120, 122, 129
Reagan administration 97
research and design institute 110, 111
research vessels 137
Riga 112
risk aversion 20
risk capital 91
risk sharing 100
Romania 112, 126

Ryzhkov 106

Sakhalin 24, 25, 37, 110, 117
Sakhalinskaya 37, 129
Salym field 22
Sattarov, Chingiz G. 54, 55
Sea of Azov 26
Sea of Okhotsk 24
Second World War 67
sector line, *see* demarcation line
seismic surveys 33, 36, 44, 46, 50, 99
semi-submersible platforms 35, 37, 128–9
service base 39
service contracts 99
Sevastopol 36
Severo-Kildinskaya 32, 33, 36, 65
Severomorsk 61
Sevmorgeo, NPO 116
Shcherbina, Boris 104, 107, 122
Shelf drilling rigs 35, 36, 128–9
Shell 47
Sheremeta, O. O. 44, 50
Shevardnadze, E. A. 106
Silayev, I. S. 122
Simferopol 112
Singapore 126
SITs Morneftegaza, *see* Soyuzmorinzhgeologiya
Skauge, Arne 121
Snorre field 49
Sodeco 24, 128
Sokolov 106
sonar range 69
South Cape 85
Soviet Far East 24
Sovmin SSSR 107
  energy policy coordination bureau 107
Soyuzmorgeo, VMNPO 44, 109
Soyuzmorinzhgeologiya, VMNPO 111, 114
  SITs Morneftegaza 112, 114
Specialized Information Centre (SITs), *see* Soyuzmorinzhgeologiya
Spetsmorneftegazproekt 110, 112
Spitsbergen, *see* Svalbard
SSBN-operations 68–82, 84
SSN-operations 82–3
state committee for construction affairs, *see* Gosstroy SSSR
state committee for hydrometeorological and environmental control, *see* Goskomgidromet SSSR
state committee for science and technology, *see* GKNT SSSR
state institute for industrial construction design, *see* Promstroyproekt
state planning committee, *see* Gosplan
Statfjord C platform 95
stationary production installations 108
strategic drilling 46
strategic stationing 58 ff.
strategic submarines 67–82 ff.
submarines 57–8, 61 ff.
substitution 12, 19
Sudoimport, V/O 36, 37, 44, 121
Sudoshelf 120
superpower politics 96
supply vessels 38, 134–5
Sushkov, V. N. 120, 121
Svalbard 26, 47, 58, 66, 67, 80, 81, 85, 97
Svalbard Treaty 85
Sweden 120

Talyzin, N. V. 106
technical assistance 94
technology imports 111
Tengiz field 23
Tenneco 47
Timano–Pechoran oil and gas province 25
Timonin, V. I. 108
Titan–1 & 2 132
trans-Atlantic relations 102
transit lines 97
Tromsøflaket 45, 46, 47
Turku 133
Typhoon 69, 76, 80
Tyumen 122

Udalov class 67
Udgaard, Nils Morten 104
Ukraine 16
Ula field 95
United Kingdom 105
United States 25, 63, 68, 71, 82, 100, 102
USS Sargo 79

Valentin Shashin 33, 35, 36
valley of death problem 73, 81
Valmet 120
Varandey field 26
vessel 53
Victor-class 67
Vietnam 37, 113
Viktor Muravlenko 33, 35, 36, 117
Vinje, Torgny 85
Volga-Ural 1
VNIIGaz 114
VNIIO 116

Vorobeva, R. S. 109
Vyborg yard 37, 56, 122

Warski yard 135
Wartsila 38, 120
water injection 21
Watkins, Admiral James D. 76, 77
West European option 102
West Germany 126
western banks 11
western equipment 91, 94
Western Europe 63, 68, 82
Western Siberia 1, 2, 20, 21–2, 27
White Sea 63
winterization 49

Yamal peninsula 26
Yastrebov, I. P. 106

Zarubezhgeologiya, VNII 114
Zaykov, L. N. 106